D0225326

Ergodic Problems of Classical Mechanics

Ergodic Problems of Classical Mechanics

V. I. ARNOLD
University of Moscow

A. AVEZ
University of Paris

This book was originally published as part of the Frontiers in Physics Series, edited by David Pines.

ADDISON-WESLEY PUBLISHING COMPANY, INC.
THE ADVANCED BOOK PROGRAM
Redwood City, California • Menlo Park, California • Reading, Massachusetts
New York • Amsterdam • Don Mills, Ontario • Sydney • Madrid
Singapore • Tokyo • San Juan • Wokingham, United Kingdom

Ergodic Problems of Classical Mechanics

Originally published in 1968 as part of The Mathematical Physics Monograph Series by W.A. Benjamin, Inc.

Library of Congress Cataloging-in-Publication Data

Arnold, V. I. (Vladimir Igorevich). 1937-
[Problèmes ergodiques de la mécanique classique. English]
Ergodic problems of classical mechanics / V. I. Arnold and A. Avez
p. cm. -- (Advanced book classics series)
Translation of: Problèmes ergodiques de la mécanique classique.
Reprint of the 1968 ed. of the translation published : New York : W. A. Benjamin.
Bibliography: p.
Includes index.
1. Dynamics. 2. Ergodic Theory I. Avez, A. (André) II. Title. III. Series.
QA845.A713 1988 531'.11--dc19 88-24096
ISBN 0-201-09406-1

Manufactured in the United States of America. Published simultaneously in Canada.

ABCDEFGHIJ-AL-89

This unique image was created with special-effects photography. Photographs of a broken road, an office building, and a rusted object were superimposed to achieve the effect of a faceted pyramid on a futuristic plain. It originally appeared in a slide show called "Fossils of the Cyborg: From the Ancient to the Future," produced by Synapse Productions, San Francisco. Because this image evokes a fusion of classicism and dynamism, the future and the past, it was chosen as the logo for the Advanced Book Classics series.

Publisher's Foreword

"Advanced Book Classics" is a reprint series which has come into being as a direct result of public demand for the individual volumes in this program. That was our initial criterion for launching the series. Additional criteria for selection of a book's inclusion in the series include:

- Its intrinsic value for the current scholarly buyer. It is not enough for the book to have some historic significance, but rather it must have a timeless quality attached to its content, as well. In a word, "uniqueness."
- The book's global appeal. A survey of our international markets revealed that readers of these volumes comprise a boundaryless, worldwide audience.
- The copyright date and imprint status of the book. Titles in the program are frequently fifteen to twenty years old. Many have gone out of print, some are about to go out of print. Our aim is to sustain the lifespan of these very special volumes.

We have devised an attractive design and trim-size for the "ABC" titles, giving the series a striking appearance, while lending the individual titles unifying identity as part of the "Advanced Book Classics" program. Since "classic" books demand a long-lasting binding, we have made them available in hardcover at an affordable price. We envision them being purchased by individuals for reference and research use, and for personal and public libraries. We also foresee their use as primary and recommended course materials for university level courses in the appropriate subject area.

The "Advanced Book Classics" program is not static. Titles will continue to be added to the series in ensuing years as works meet the criteria for inclusion which we've imposed. As the series grows, we naturally anticipate our book buying audience to grow with it. We welcome your support and your suggestions concerning future volumes in the program and invite you to communicate directly with us.

Advanced Book Classics

1989 Reissues

V.I. Arnold and A. Avez, *Ergodic Problems of Classical Mechanics*

E. Artin and J. Tate, *Class Field Theory*

Michael F. Atiyah, *K-Theory*

David Bohm, *The Special Theory of Relativity*

Ronald C. Davidson, *Theory of Nonneutral Plasmas*

P.G. de Gennes, *Superconductivity of Metals and Alloys*

Bernard d'Espagnat, *Conceptual Foundations of Quantum Mechanics, 2nd Edition*

Richard Feynman, *Photon-Hadron Interactions*

William Fulton, *Algebraic Curves: An Introduction to Algebraic Geometry*

Kurt Gottfried, *Quantum Mechanics*

Leo Kadanoff and Gordon Baym, *Quantum Statistical Mechanics*

I.M. Khalatnikov, *An Introduction to the Theory of Superfluidity*

George W. Mackey, *Unitary Group Representations in Physics, Probability and Number Theory*

A. B. Migdal, *Qualitative Methods in Quantum Theory*

Phillipe Nozières and David Pines, *The Theory of Quantum Liquids, Volume II* - new material, 1989 copyright

David Pines and Phillipe Nozières, *The Theory of Quantum Liquids, Volume I: Normal Fermi Liquids*

David Ruelle, *Statistical Mechanics: Rigorous Results*

Julian Schwinger, *Particles, Source and Fields, Volume I*

Julian Schwinger, *Particles, Sources and Fields, Volume II*

Julian Schwinger, *Particles, Sources and Fields, Volume III* - new material, 1989 copyright

Jean-Pierre Serre, *Abelian ℓ-Adic Representations and Elliptic Curves*

R.F. Streater and A.S. Wightman, *PCT Spin and Statistics and All That*

René Thom, *Structural Stability and Morphogenesis*

Vita

Vladimir I. Arnold
Professor of Mathematics at Moscow University and at the Steklov Mathematics Institute.
A graduate of Moscow University, Dr. Arnold received his Ph.D. in Mathematical Science in 1965. Dr. Arnold is the author of six books and is a member of the London Mathematical Society, the National Academy of Sciences, the Academie des Sciences, the Academy of Sciences U.S.S.R., the American Academy of Arts and Sciences and the Berlin Academy of Sciences. He is also the recipient of the Stalin Prize in Mathematics.

André Avez
Professor of Mathematics at the University of Paris, where he received his Ph.D. in Mathematics in 1963. He was Associate Professor of Mathematics at the University of Minnesota from 1965-1966. Dr. Avez's main research interests include the study of general relativity, topology and analysis on manifolds, and dynamical systems.

Special Preface

Since this book was first written, ergodic theory has become very popular and there are many new expositions. We note, for instance, "Ergodic Theory" by I.P. Cornfeld, S.V. Fomin, Ya. G. Sinai, "Mathematical Methods of Classical Mechanics" and "Geometrical Methods of the Theory of Ordinary Differential Equations" by V.I. Arnold, also "Ergodic Theory and Differentiable Dynamics" by R. Mañé. See also the 6 volumes (the dynamical systems section) of the Encyclopedia of Mathematical Sciences (Springer 1988 translation of Moscow VINITI Series "Itogi nauki, Fundamentalnye napravlenia." Moscow, 1985). In addition, averaging theory (works by N.N. Nehoroshev, A.I. Neistadt and V.I. Bahtin) connects important new results with the contents of the present volume. Many numerical experiments exist on ergodic properties of complicated systems, including the hydrodynamical ones, but practically no new theoretical results have been discovered in this domain. Therefore, we feel that the results of this book are still valid and useful to those interested in ergodic problems.

Preface

The fundamental problem of mechanics is computing, or studying qualitatively, the evolutions of a dynamical system with prescribed initial data.

Numerical methods allow one to compute the orbits for a final time interval, but they fail as the time increases indefinitely. The three-body problem offers a typical example: Do there exist arbitrarily small perturbations of the initial data for which one of the bodies moves to infinity? Mathematically speaking, the problem is the study of the orbits of a vector field on phase-space. Far from being solved, such a problem involves areas as various as probability and topology, number theory and differential geometry. Mr. Nicholas Bourbaki may forgive us for mixing so many structures.

Maxwell, Boltzmann, Gibbs and Poincaré first proposed a statistical study of complex dynamical systems, which is now known as ergodic theory. [Ergodic theory was conceived for mechanics but applied to various other branches, such as number theory. For example, how are the first digits 1, 2, 4, 8, 1, 3, 6,...of the powers 2^n distributed? (*See* Appendix 12.)] But the mathematical definitions and the first important theorems are due to J. von Neumann, G.D. Birkhoff, E. Hopf, and P.R. Halmos, and they appeared only in the Thirties. During the past decade, a new step was taken, inspired by Shannon's information theory. The main result, due to Kolmogorov, Rohlin, Sinai, and Anosov, consists in a deep study of a strongly stochastic class of dynamical systems. This class is wide enough to include all the sufficiently unstable classical systems. Among these systems figure the geodesic flows of space with negative curvature, as studied by Hadamard, Morse, Hedlund, E. Hopf, Gelfand, and Fomin. On the other hand, Sinai proved that the Boltzmann-Gibbs model, that is, a system of hard spheres with elastic collisions, belongs also to this class; this proved the "ergodic conjecture."

This book is by no means a complete treatise on ergodic theory, and references are not exhaustive.

The text presented here is based on lectures delivered during the spring and fall of 1965 by one of the authors, who also wrote Chapter 4. The second author is responsible for the proofs of Chapters 1, 2 and 3.

We thank Professors Y. Choquet-Bruhat, H. Cabannes and P. Germain, J. Kovalewsky, G. Reeb, L. Schwartz, R. Thom and M. Zerner, who welcomed the lecturer at their seminar. We also thank Professor S. Mandelbrojt, who suggested that we write this book. The final manuscript was read by Y. Sinai, who made a number of useful improvements for which we are sincerely grateful.

The translator (A. Avez) wishes to thank warmly Professors V.I. Arnold, S. Deser, and A.S. Wightman, who prevented him from many mistakes.

V.I. Arnold
A. Avez

Contents

CHAPTER 1

DYNAMICAL SYSTEMS

This chapter contains examples of dynamical systems and related problems.

§1. Classical Systems

DEFINITION 1.1

Let M be a smooth manifold, μ a measure on M defined by a continuous positive density, ϕ_t: $M \to M$ a one-parameter group of measure-preserving diffeomorphisms. The collection (M, μ, ϕ_t) is called a classical dynamical system.

The parameter t is a real number or an integer. If $t \in \mathbf{R}$, the group ϕ_t is usually defined in local coordinates by:

$$\dot{x}^i = f^i(x^1, \ldots, x^n), \quad i = 1, \ldots, \quad n = \dim M.$$

If $t \in \mathbf{Z}$, ϕ_t is the discrete group generated by a measure-preserving diffeomorphism $\phi = \phi_1$. Then the system is merely denoted by (M, μ, ϕ) and ϕ is called the automorphism.

EXAMPLE 1.2. QUASI-PERIODIC MOTION

Let M be the torus $\{(x, y) \bmod 1\}$. The measure is $dx\,dy$, the group ϕ_t is a translation group:

$$\dot{x} = 1, \quad \dot{y} = \alpha$$

where $\alpha \in \mathbf{R}$, and dot denotes d/dt. Assume $\alpha = p/q$ rational:

$$p, q \in \mathbf{Z}, \quad q > 0$$

and p and q relatively prime. In the covering plane (x, y), the orbit with initial data $x(0) = x_0$, $y(0) = y_0$ has the form:

$$y = y_0 + \frac{p}{q}(x - x_0) .$$

As $x = x_0 + q$, y takes the value $y_0 + p$ and the corresponding point on M coincides with the initial point (x_0, y_0). Thus, the torus is covered by closed orbits. If α is irrational, each orbit is everywhere dense (Jacobi, 1835; see Appendix 1). More generally, let $T^n = \{(x^1, \ldots, x^n) \bmod 1\}$ be the n-dimensional torus with the usual measure $dx^1 \cdots dx^n$, and ϕ_t the

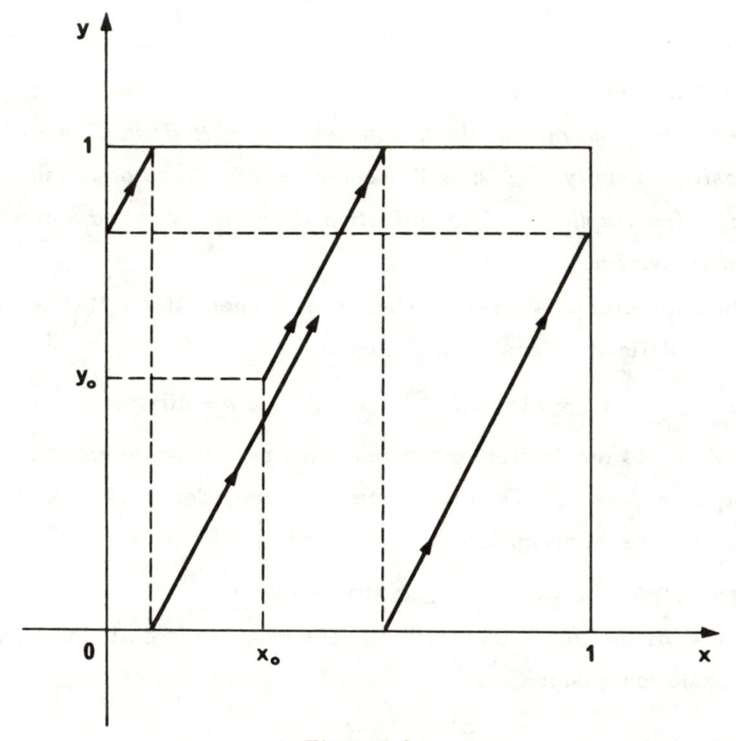

Figure 1.3

one-parameter group of translations defined by:

$$\dot{x}^i = \omega^i \; ; \quad i = 1, \ldots, n ; \quad \omega \in R^n .$$

Every orbit of ϕ_t is everywhere dense if, and only if,

$$k \in Z^n \text{ and } \omega \cdot k = 0 \text{ imply } k = 0 .$$

EXAMPLE 1.4. GEODESIC FLOWS

Let V be a compact Riemannian manifold; $M = T_1V$ denotes its unitary tangent bundle. Given a unit tangent vector $\xi \in T_1V_x$ to V at x, there is one, and only one, geodesic γ passing through x with initial velocity vector ξ. We denote by $\gamma(\xi, s)$ the point of γ obtained from x in time s when moving along γ with velocity 1. The unit tangent vector to γ at $\gamma(\xi, s)$ is

$$G_t\xi = \frac{d}{ds}\gamma(\xi, s)\Big|_{s=t} \in T_1V_{\gamma(\xi,t)} \quad . \tag{1.5}$$

Formula (1.5) defines a one-parameter group of diffeomorphisms of $M = T_1V$.

DEFINITION 1.6

The group G_t *is called the geodesic flow of* V. It can be proved that G_t preserves the measure μ induced on M by the Riemannian metric of V (Liouville's theorem).

SOME MORE EXAMPLES 1.7

Appendix 2 describes the geodesic flow of the usual torus immersed in the Euclidean space E^3. For the ellipsoid see Kagan [1], and for Lie groups with a left-invariant metric see Appendixes 3 and 4. One more word, in mechanics geodesic flow is called "movement of a material point on a frictionless surface without external forces." Other mechanical systems involve more general flows.

EXAMPLE 1.8. HAMILTONIAN FLOWS

Let $p_1, \ldots, p_n$; $q_1, \ldots, q_n$ (in short: p, q) be a coordinate system in $\mathbf{R}^{2n}$, and $H(p, q)$ a smooth function. The equations

$$\frac{dq}{dt} = \frac{\partial H}{\partial p}, \quad \frac{dp}{dt} = -\frac{\partial H}{\partial q} \tag{1.9}$$

define a one-parameter group of diffeomorphisms of $\mathbf{R}^{2n}$. This group is called a Hamiltonian flow on $\mathbf{R}^{2n}$.

LIOUVILLE'S THEOREM 1.10

The Hamiltonian flow preserves the measure $dp_1 \cdots dp_n \cdot dq_1 \cdots dq_n$.

Proof:

The divergence of the vector field (1.9) vanishes:

$$\frac{\partial}{\partial q}\left(\frac{\partial H}{\partial p}\right) + \frac{\partial}{\partial p}\left(-\frac{\partial H}{\partial q}\right) = 0 \ . \qquad \text{(Q. E. D.)}$$

THEOREM OF CONSERVATION OF ENERGY 1.11

The function H *is a first integral of* (1.9).

Proof:

$$\frac{dH}{dt} = \frac{\partial H}{\partial q} \cdot \frac{\partial H}{\partial p} + \frac{\partial H}{\partial p} \cdot \left(-\frac{\partial H}{\partial q}\right) = 0 \qquad \text{(Q. E. D.)}$$

Let us denote a subset $H(p, q) = h \in R$ by M. For almost every h, M is a manifold. This manifold is invariant under the flow.

COROLLARY 1.12

There exists an invariant measure on the manifold M.

Proof:

The invariant measure on M is defined by:

$$d\mu = \frac{d\sigma}{\|\mathbf{grad}\ H\|}, \quad \|\ \ \| = \text{length},$$

where σ is the volume element of M induced by the metric of R^{2n}. If (1.9) has several first integrals, namely $I_1, I_2, \ldots, I_k$, then the equations (1.9) determine a classical dynamical system on each $(2n-k)$-dimensional manifold: $I_1 = h_1, \ldots, I_k = h_k$, where the h's are constants.

EXAMPLE 1.13. LINEAR OSCILLATIONS IN DIMENSION 2

The Hamiltonian is:

$$H = \frac{\omega 1}{2}(p_1^2 + q_1^2) + \frac{\omega 2}{2}(p_2^2 + q_2^2) \ .$$

Equation (1.9) has two first integrals:

$$I_1 = p_1^2 + q_1^2, \qquad I_2 = p_2^2 + q_2^2 \ .$$

The corresponding manifolds $I_1 = h_1$, $I_2 = h_2$ are two-dimensional tori.

The dynamical systems that are induced on these tori are isomorphic to those of Example (1.2). Appendix 5 provides further examples.

REMARK 1.14. GLOBAL HAMILTONIAN FLOWS

More generally one may consider a symplectic [1] $2n$-dimensional manifold M^{2n} instead of R^{2n}, and a closed one-form ω_1 $(= dH)$ instead of H. Equation (1.9) becomes

$$\dot{x} = I\omega_1, \quad x \in M^{2n}$$

where $I: T^*M_x \to TM_x$ is defined by

$$\Omega(I\omega_1, \xi) = \omega_1(\xi),$$

for any $\xi \in TM_x$. Let us now give some examples of the discrete case: $t \in Z$.

EXAMPLE 1.15. TRANSLATIONS OF THE TORUS

Let M be the torus $\{(x, y) \bmod 1\}$ with the usual measure $dx\,dy$. The automorphism ϕ is

$$\phi(x, y) = (x + \omega_1, y + \omega_2) \pmod 1, \qquad \omega_1 \in R.$$

Each orbit of ϕ is everywhere dense if, and only if, $k \cdot \omega \in Z$, for $k \in Z$, imply $k = 0$ (see Appendix 1).

EXAMPLE 1.16. AUTOMORPHISMS OF THE TORUS

Again $M = \{(x, y) \bmod 1\}$ and $d\mu = dx\,dy$. The automorphism ϕ is defined by:

$$\phi(x, y) = (x + y, x + 2y) \pmod 1 .$$

The mapping ϕ induces a linear mapping in the covering plane (x, y)

$$\tilde{\phi} = \begin{pmatrix} 1 & 1 \\ 1 & 2 \end{pmatrix}.$$

As Det $\tilde{\phi} = 1$, ϕ is measure-preserving. A set A is transformed under

---

[1] A symplectic manifold M^{2n} is a smooth manifold, together with a global closed two-form Ω of rank n. Example:

$$\Omega = dp \wedge dq \quad \text{on } R^{2n} .$$

$\tilde{\phi}$, and then $\tilde{\phi}^2$ as pictured in Figure (1.17). The linear mapping $\tilde{\phi}$ has two real proper values λ_1 and λ_2: $0 < \lambda_2 < 1 < \lambda_1$.

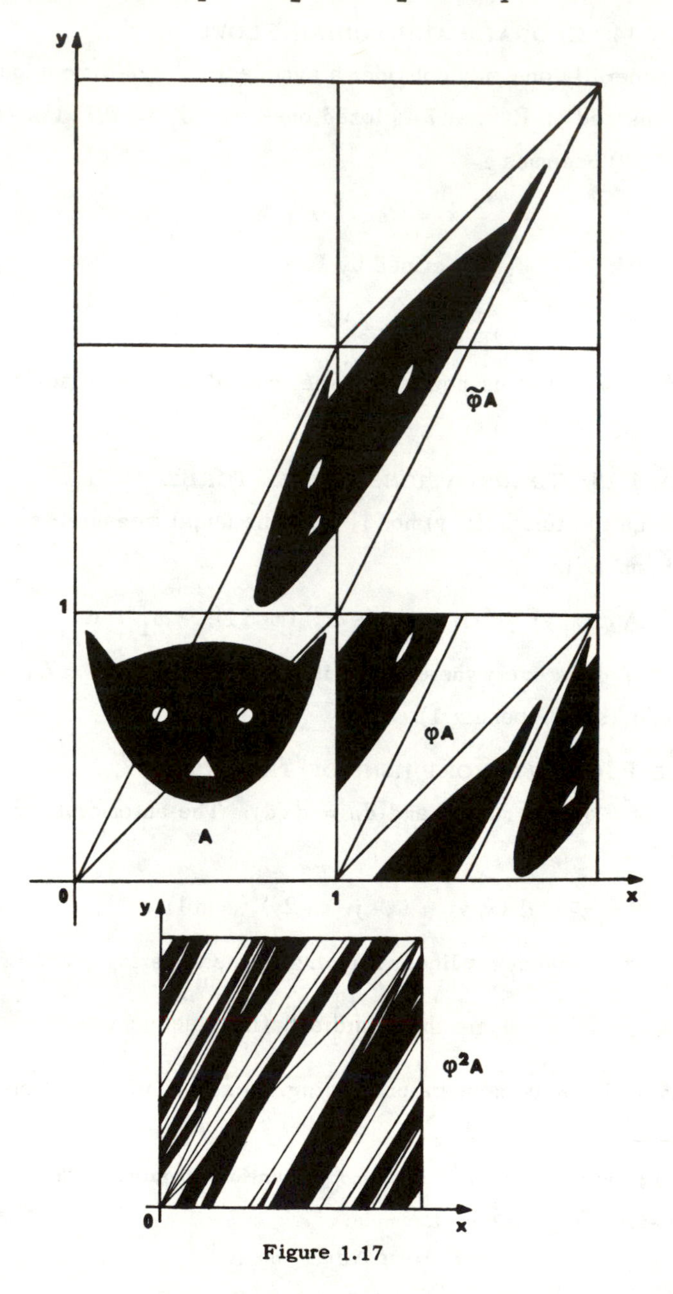

Figure 1.17

Then, for n large enough, $\tilde{\phi}^n A$ looks like a very long and very narrow ribbon of the plane. On M, this ribbon lies approximately in the neighborhood of an orbit of the system:

$$\dot{x} = 1, \qquad \dot{y} = \lambda_1 - 1 :$$

According to Jacobi's theorem (Example 1.2), and because $\lambda_1 - 1$ is irrational, $\phi^n A$ converges to a dense helix of the torus as $n \to +\infty$.

§2. Abstract Dynamical Systems

DEFINITION 2.1[2]

An abstract dynamical system (M, μ, ϕ_t) is a measure-space (M, μ) *equipped with a one-parameter group* ϕ_t *of automorphisms* (mod 0) *of* (M, μ), ϕ_t *depending measurably of* t.

Thus, for any measurable sets A and B, $\mu(\phi_t A \cap B)$ is a measurable function of t, and $\mu(\phi_t A) = \mu(A)$ for any t. In the future (M, μ) will always be a nonatomic Lebesgue space, that is (M, μ) will be isomorphic modulo 0 to $[0, 1]$ with its usual Lebesgue measure. In particular $\mu(M) = 1$.

If ϕ_t is the discrete group generated by an automorphism $\phi = \phi_1$, we merely denote the system by (M, μ, ϕ). In the following we shall omit the notation "mod 0." All of the preceding examples are abstract systems: a compact Riemannian manifold M with its canonical measure μ $(\mu(M) = 1)$ is isomorphic to $[0, 1]$.

EXAMPLE 2.2. BERNOUILLI SCHEMES

The space M. Let $\mathbf{Z}_n = \{0, 1, \ldots, n-1\}$ be the first n nonnegative integers. M is the Cartesian product $M = \mathbf{Z}_n^{\mathbf{Z}}$ of a countable family of $\mathbf{Z}_n$'s. Thus, the elements m of M are the bilateral infinite sequences of elements of $\mathbf{Z}_n$:

$$m \in M, \qquad m = \cdots a_{-1}, a_0, a_1, \ldots .$$

---

[2] See Appendix 6 for these concepts.

The σ-algebra of the measurable sets. It is the algebra generated by sets of the form

$$A_i^j = \{m \mid a_i = j\}, \quad i \in \mathbf{Z}, \ j \in \mathbf{Z}_n .$$

The measure μ. Define a normalized measure μ on $\mathbf{Z}_n$ by setting:

$$\mu(0) = p_0, \ldots, \mu(n-1) = p_{n-1}, \ \Sigma p_i = 1 .$$

We set $\mu(A_i^j) = p_j$ for every i, j. The measure of σ is the product-measure, denoted again by μ: if $A_{i_1}^{j_1}, \ldots, A_{i_k}^{j_k}$ ($i_1, \ldots, i_k$ all different) are k-distinct generators, the measure of their intersection is the product of their measures, that is,

$$\mu\{m \mid a_{i_1} = j_1, \ldots, a_{i_k} = j_k\} = p_{j_1} \cdots p_{j_k} .$$

(M, μ) is clearly a Lebesgue space.

The automorphism ϕ. It is the shift

$$m = (\ldots, a_i, \ldots) \to m' = (\ldots, a_i', \ldots) ,$$

where $a_i' = a_{i-1}$ for every i; ϕ is a bijection. To prove ϕ is measure-preserving, it is sufficient to take into account the generators:

$$\phi(A_i^j) = \{\phi(m) \mid a_i = j\} = \{m' \mid a_{i+1}' = j\} = A_{i+1}^j .$$

Hence:

$$\mu[\phi(A_i^j)] = \mu[A_{i+1}^j] = p_j = \mu(A_i^j).$$

Notation. The above abstract dynamical system is called a Bernouilli scheme and denoted by $B(p_0, \ldots, p_{n-1})$.

Remark. Tossing a coin involves the scheme $B(\frac{1}{2}, \frac{1}{2})$. This fact was first pointed out by J. Bernouilli. The elements of $M = \mathbf{Z}_2^{\mathbf{Z}}$ are indefinite bilateral sequences of tosses: 0 means "head," 1 means "tail." The set A_i^0 (resp. A_i^1) represents the set of the sequence in which "head" (resp. "tail") appears at the ith toss. Thus, it is quite natural to set:

$$\mu(A_i^j) = \text{prob}(A_i^j) = \frac{1}{2} .$$

EXAMPLE 2.3. THE BAKER'S TRANSFORMATION

Let M be the torus $\{(x, y) \bmod 1\}$ with its usual measure $dx\,dy$.

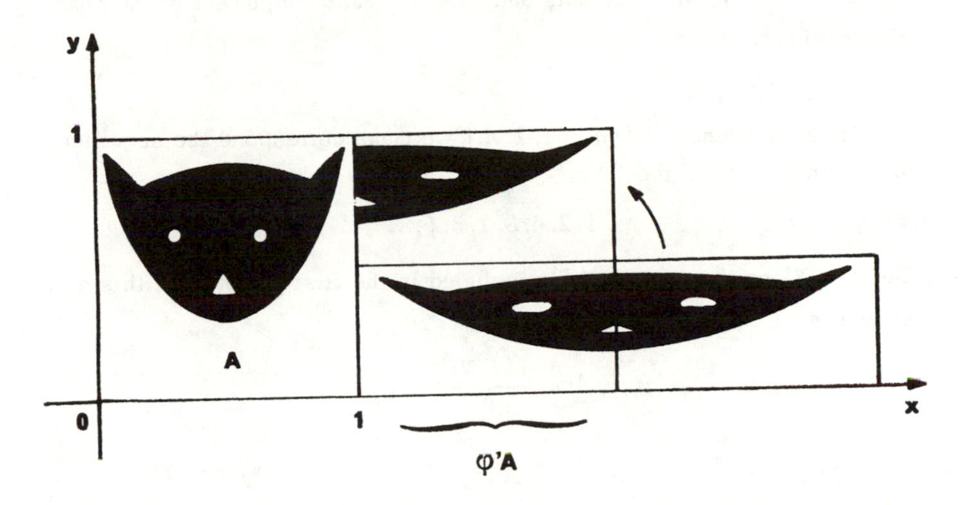

Figure 2.4

The automorphism ϕ' is defined as:

$$\phi'(x, y) = \begin{cases} (2x, \tfrac{1}{2}y) & \text{mod } 1, \text{ if } 0 \le x < \tfrac{1}{2} \\ (2x, \tfrac{1}{2}(y+1)) & \text{mod } 1, \text{ if } \tfrac{1}{2} \le x < 1 . \end{cases}$$

To study ϕ', it is convenient to introduce the induced mapping $\tilde{\phi}'$ in the covering plane (x, y); $\tilde{\phi}'$ can be described as follows: Transform the unit square by the affinity, making it twice as long as before in the direction ox, twice as short in the direction oy. Then, cut off the right half of this rectangle and move it, by translation, above the left half (see Figure 2.4). As $A \subset M$ and n converges to $+\infty$, $\phi'^n A$ is formed by a "very large number" of segments which are parallel to ox.

Let us mention some problems on dynamical systems.

§3. Computations of Mean Values

EXAMPLE 3.1

Lagrange [1], studying the problem of mean motion of the perihelion, raised the following question: compute, if it exists

$$\lim_{t \to \infty} \frac{1}{t} \operatorname{Arg} \sum_{k=1}^{3} a_k \cdot \exp(i\omega_k t)$$

where the a_k, ω_k are constants and Arg z means "argument of the complex number z."

EXAMPLE 3.2

To the sequence $\{2^n \mid n = 1, 2, \ldots\}$ we make correspond the sequence of the first digits of the 2^n's:

$$1, 2, 4, 8, 1, 3, 6, \ldots .$$

Let $\tau(7, N)$ be the number of 7's included in the first N terms of this new sequence. Compute (if it exists)

$$\lim_{N \to \infty} \frac{\tau(7, N)}{N} = p_7 .$$

EXAMPLE 3.3

Let D be a region of a Riemannian space, and $\gamma(t)$ a geodesic. What is the mean sojourn time of $\gamma(t)$ in D? In other words, we set

$$\tau(T) = \text{measure } \{t \mid 0 \leq t \leq T, \ \gamma(t) \in D\}$$

and we ask for

$$\lim_{T \to \infty} \frac{\tau(T)}{T} .$$

These three problems are special cases of a more general one: Let f be a complex-valued, μ-summable function defined on the space M of an abstract system (M, μ, ϕ_t). Compute (if it exists)

$$\lim_{T \to \infty} \frac{1}{T} \int_0^T f(\phi_t m) dt, \qquad m \in M.$$

In (3.3), f is none other than the characteristic function of $T_1 D$. Of course, there exist more problems involving some computations of means:

EXAMPLE 3.4

Let (M, μ, ϕ_t) be a dynamical system, and A and B two measurable sets of M. Compute (if it exists)

$$\lim_{t \to +\infty} \mu[\phi_t A \cap B]$$

(see Figures 1.17 and 2.4). Physical intuition makes it plausible that for a probabilistic "enough" system, the limit exists and is equal to

$$\mu(A) \cdot \mu(B) .$$

§4. Problems of Classification

Isomorphism of Abstract Dynamical Systems

A natural way to classify the dynamical systems is to exhibit their invariants with respect to their corresponding group: canonical transformations for Hamiltonian flows, measure-preserving diffeomorphisms for general classical flows. Since abstract invariants are the deepest ones, we give the definition:

DEFINITION 4.1

Two abstract dynamical systems (M, μ, ϕ) and (M', μ', ϕ') are isomorphic if there exists an isomorphism $f: M \to M$ (mod 0) of measurable spaces making the following diagram commutative:

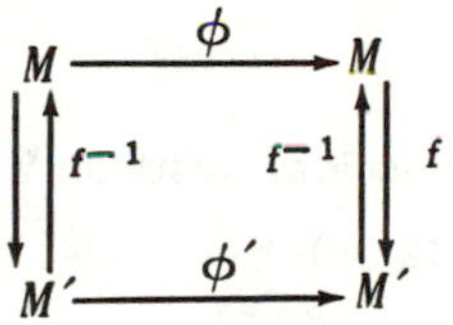

$$\begin{array}{ccc} M & \xrightarrow{\phi} & M \\ f \big\downarrow \big\uparrow f^{-1} & & f^{-1} \big\uparrow \big\downarrow f \\ M' & \xrightarrow{\phi'} & M' \end{array}$$

A similar definition holds in the continuous case.

EXAMPLE 4.2

The Bernouilli schemes $B(1/2, 1/8, 1/8, 1/8, 1/8)$ and $B(1/4, 1/4, 1/4, 1/4)$ are isomorphic (see Meshalkin [1], Blum and Hanson [1]).

EXAMPLE 4.3

The translations of the tori (1.15) are not isomorphic to the automorphism of the torus (1.16) (see Chapter 2, 12.40).

EXAMPLE 4.4

On the torus $T^2 = \{(x, y) \bmod 1\}$ with the usual measure, let us consider the automorphisms:

$$\phi(x, y) = (3x+y, 2x+y) \pmod 1; \qquad \phi'(x, y) = (3x+2y, x+y) \pmod 1 .$$

Both of them are nonisomorphic to the automorphism (1.16) (see Corollary 12.30), but their isomorphism is still an open question.

EXAMPLE 4.5

The Bernouilli scheme $B(\frac{1}{2}, \frac{1}{2})$ is isomorphic to the Baker's transform (see Appendix 7 for a proof). One problem among the fundamental problems of Ergodic Theory is to find necessary and sufficient conditions under which two Bernouilli schemes are isomorphic.

§5. Problems of Generic Cases

Faced with such a diversity of dynamical systems, it seems useful to clarify the situation by neglecting the "exceptional cases." The word "exceptional" becomes meaningful by putting a topology or a measure over the group of the automorphisms. A class of dynamical systems can be exceptional in the abstract frame, and generic in the classical one, or conversely.

EXAMPLE 5.1

There exist abstract dynamical systems which are nonisomorphic to any classical system (see 12.39).

EXAMPLE 5.2

In the abstract frame, the mixing is exceptional in the weak topology (see Halmos [1], Rohlin [1]). By contrast, every diffeomorphism, C^1- close enough to the automorphism ϕ: $T^2 \to T^2$ of Example (1.16), is mixing. Thus, mixing can be generic in the classical frame.

EXAMPLE 5.3

In the abstract frame, ergodic systems are generic in the weak topology (see Halmos [1]). By contrast, every Hamiltonian system close enough to the geodesic flow of the torus T^2 (see Appendix 2) is nonergodic. See also the three-body system (Chapter 4). Thus, ergodicity can be nongeneric in the classical frame.

General References for Chapter 1

Abraham, R., *Foundations of Mechanics*, Benjamin (1967).

Birkhoff, G. D., *Dynamical Systems*, American Mathematical Society Colloquium Publications 9, New York (1927).

Godbillon, C., *Géométrie différentielle et mécanique*, Hermann, Paris (1968).

Halmos, P. R., *Measure Theory*, Chelsea, New York (1958).

Halmos, P. R., *Lectures on Ergodic Theory*, Chelsea, New York (1959).

Whittaker, E. T., *Analytical Dynamics*, Dover, New York (1944).

CHAPTER 2

ERGODIC PROPERTIES

A series of concepts (ergodicity, mixing, spectrum, entropy, etc.) has been introduced by the metric theory of dynamical systems to describe the behavior of most of the orbits. This chapter is devoted to their definition. In Chapters 3 and 4 we shall make use of these concepts to describe classical systems.

§6. Time Mean and Space Mean

DEFINITION 6.1. TIME MEAN

Let (M, μ, ϕ) or (M, μ, ϕ_t) be a dynamical system, f a complex-valued function defined on M. The time mean $\overset{*}{f}$ of f, if it exists, is defined by:

$$(6.2) \qquad \overset{*}{f}(x) = \lim_{N \to +\infty} \frac{1}{N} \sum_{n=0}^{N-1} f(\phi^n x), \qquad x \in M, \quad n \in \mathbf{Z}^+$$

in the discrete case, and by:

$$(6.2)' \qquad \overset{*}{f}(x) = \lim_{T \to +\infty} \frac{1}{T} \int_0^T f(\phi_t x) dt, \qquad x \in M, \quad t \in \mathbf{R}$$

in the continuous case.

DEFINITION 6.3. SPACE MEAN

It is defined, if it exists, as:

$$\bar{f} = \int_M f(x) d\mu .$$

(Recall that $\mu(M) = 1$.)

THEOREM 6.4. (G. D. BIRKHOFF – A. J. KHINCHIN)[1]

Let (M, μ, ϕ_t) be an abstract dynamical system, $f \in L_1(M, \mu)$ a complex-valued μ-summable function on M. Then:

(a) $\overset{*}{f}(x)$ exists almost everywhere (abbreviation a.e.), that is except, perhaps, on a set of measure zero;

(b) $\overset{*}{f}(x)$ is summable and invariant a.e., that is:

$$\overset{*}{f}(\phi_t x) = \overset{*}{f}(x) \quad \text{for every } t,$$

except, perhaps, on a set of measure zero independent of t;

(c) $$\int_M \overset{*}{f}(x) d\mu = \int_M f(x) d\mu .$$

A proof will be found in P. Halmos [1] for the discrete case, in Nemytskii-Stepanov [1] for the continuous case.

REMARK 6.5

Examples of systems (M, μ, ϕ_t) and functions f can be found, in which $\overset{*}{f}(x)$ does not exist, or is not equal to $\overline{f}$, on a dense subset of M; even if f is analytic and (M, μ, ϕ_t) classical (see Appendix 8).

REMARK 6.6

There exist dynamical systems in which $\overset{*}{f}$ exists everywhere as soon as f is continuous or even Riemannian-integrable. For instance, the translations of the torus (see Examples 1.2 and 1.15, Appendix 9).

§7. Ergodicity

DEFINITION 7.1

An abstract dynamical system is ergodic if for every complex-valued μ-summable function $f \in L_1(M, \mu)$ the time mean is equal to the space mean a.e.:

$$\overset{*}{f}(x) = \overline{f(x)}, \quad \text{a.e.} \tag{7.2}$$

---

[1] Appendix 10 provides some applications of this theorem to Differential Geometry.

Thus, for an ergodic system, the time mean does not depend on the initial point x.

EXAMPLE 7.3

Let us assume that M is the disjoint union of two sets M_1 and M_2 of positive measure, each of which is invariant under ϕ (see Figure 7.4):

$$\phi M_1 = M_1, \quad \phi M_2 = M_2 .$$

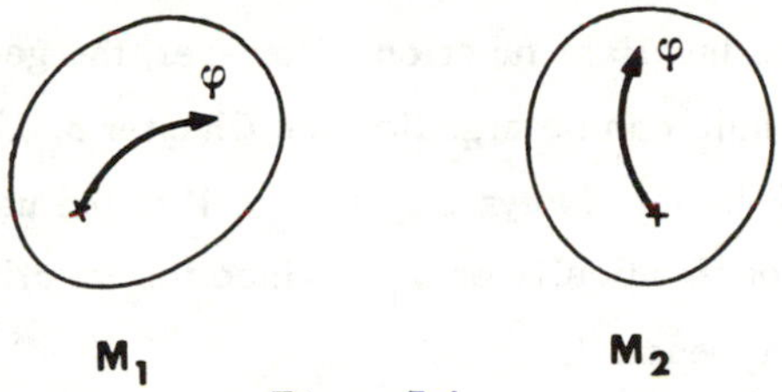

Figure 7.4

Such a system (M, μ, ϕ) is called decomposable. A decomposable system is not ergodic. In fact, taking

$$f(x) = \begin{cases} 1 & \text{if } x \in M_1 \\ 0 & \text{if } x \in M_2 , \end{cases}$$

the time mean $\overset{*}{f}(x) = f(x)$ depends on x.

REMARK 7.5

Conversely, a nonergodic system (M, μ, ϕ) is decomposable. In fact, nonergodicity implies there exists a function $f \in L_1(M, \mu)$, the time mean of which is not constant a.e. (this function can be assumed real-valued: take $\mathcal{R}f$ or $\mathcal{I}f$). Set:

$$M_1 = \{x \mid \overset{*}{f}(x) < a\}, \quad M_2 = \{x \mid \overset{*}{f}(x) \geq a\} .$$

For a suitable a we get:

$$\mu(M_1) > 0, \quad \mu(M_2) > 0 .$$

According to the Birkhoff theorem the time mean is invariant under ϕ, hence:

$$\phi M_1 = M_1, \quad \phi M_2 = M_2$$

and the system is decomposable. Whence:

COROLLARY 7.6

An abstract dynamical system is ergodic if, and only if, it is indecomposable, that is if every invariant measurable set has measure 0 *or* 1.

The preceding argument proves more: a system is ergodic if, and only if, any invariant measurable function $f \in L_1(M, \mu)$ is constant a.e.

EXAMPLE 7.7

Hamiltonian flows (Chapter 1, Theorem 1.11) are never ergodic since the energy H is an invariant function. However, the geodesic flow on the unitary tangent bundle can be ergodic (see Chapter 3, 17.12). But the geodesic flow on T_1V is not always ergodic: if V is the usual torus, the geodesic flow acts nonergodically on T_1V, since the function $\dot{\phi}(1 + r \cos \psi)^2$ is invariant (see Appendix 2).[2]

EXAMPLE 7.8

The rotation $\phi: x \to x + a \pmod 1$ *of the circle* $M = \{x \pmod 1\}$ *is ergodic if, and only if,* a *is not rational.*

Proof:

First case, a is rational. We set $a = p/q$, $p, q \in Z$, $q > 0$, p and q relatively prime. Since $f(x) = e^{2\pi i q x}$ is a nonconstant measurable invariant function, ϕ is not ergodic.

Second case, a is nonrational. Let A be an invariant set of positive measure; we shall prove $\mu(M) = 1$. Since $\mu(A) > 0$, A has a density point x_0 (Lebesgue), that is for any ε with $0 < \varepsilon < 1$ there is an arc $I =]x_0 - \delta, x_0 + \delta[$ of length at most ε, such that

$$\mu(A \cap I) \geq (1 - \varepsilon)\mu(I).$$

From the invariance of A and μ, we obtain:

$$\mu(A \cap \phi^n I) \geq (1 - \varepsilon) \cdot \mu(\phi^n I) .$$

Thus, if $n_1, \ldots, n_k$ are integers for which $\phi^{n_1}I, \ldots, \phi^{n_k}I$ are disjoint, we get:

---

[2] Small analytic perturbations of this metric of T^2 preserve the nonergodicity of the geodesic flow (see Chapter 4).

$$\mu(A) \geq \sum_{i=1}^{k} \mu(A \cap \phi^{n_i} I) \geq (1-\varepsilon)\mu\left(\bigcup_{i=1}^{k} \phi^{n_i} I\right) .$$

On the other hand, the orbit of a given endpoint of I is dense (Jacobi theorem, Appendix 1). Since $\mu(I) \leq \varepsilon$, there exist integers $n_1,\ldots, n_k$ such that the sets $\phi^{n_1} I,\ldots, \phi^{n_k} I$ are disjoint and cover M up to a set of measure 2ε. Consequently

$$\mu\left(\bigcup_{i=1}^{k} \phi^{n_i} I\right) \geq 1-2\varepsilon$$

and

$$\mu(A) \geq (1-\varepsilon)(1-2\varepsilon) .$$

Since ε is arbitrary, $\mu(A) = 1$ and the system is ergodic. A similar argument proves that the systems of Examples (1.2) and (1.15) are ergodic as soon as their orbits are everywhere dense (see Appendix 11). Appendixes 12 and 13 provide more examples.

§8. Mixing

Let M be a shaker full of an incompressible fluid, which consists of 20% rum and 80% Coca Cola (see Figure 8.1). If A is the region originally occupied by the rum, then, for any part B of the shaker, the percentage of rum in B, after n repetitions of the act of stirring, is

$$\frac{\mu(\phi^n A \cap B)}{\mu(A)} .$$

In such a situation, physicists expect that, after the liquid has been stirred sufficiently often ($n \to +\infty$), every part B of the shaker will contain approximately 20% rum. This leads to the following definition.

Figure 8.1

DEFINITION 8.2

An abstract dynamical system (M, μ, ϕ_t) *is mixing if:*

$$\lim_{t \to +\infty} \mu[\phi_t A \cap B] = \mu(A) \cdot \mu(B) \tag{8.3}$$

for every pair of measurable sets A, B.

It is clear that a dynamical system isomorphic to a mixing one is mixing. Thus, mixing is an invariant property of the dynamical systems.

COROLLARY 8.4

Mixing implies ergodicity.

Proof:

Let A be an invariant measurable set. Take $B = A$; we get:

$$\phi_t A \cap A = A$$

and from (8.3):

$$\mu(A) = 0 \text{ or } 1.$$

The following example proves the converse is false: ergodicity does not imply mixing.[3]

EXAMPLE 8.5

An isometry ϕ of a Riemannian manifold is never mixing because the images $\phi^n A$ of a small set A are congruent to A, and so their intersection with another set B is sometimes empty, sometimes of positive measure. For instance, the ergodic translations of the tori (Examples 1.2 and 1.15) cannot be mixing.

EXAMPLE 8.6

Comparison of Figures (1.17), (2.4), and (8.1) suggests a conjecture: the automorphism (1.16) of the torus T^2 and the Bernouilli schemes are mixing. This will be proved in (10.5) and (10.6).

REMARK 8.7

Mixing can be defined for endomorphisms (Appendix 6) which are not automorphisms (see Appendix 14).

REMARK 8.8

Between ergodicity and mixing there is another concept, which is also an invariant of the dynamical systems, the concept of weak mixing (see Halmos [1]). A dynamical system (M, μ, ϕ_t) is, by definition, weakly mixing if

$$\lim_{T\to+\infty} \frac{1}{T}\int_0^T |\mu(\phi_t A\cap B) - \mu(A)\cdot\mu(B)|\,dt = 0$$

in the continuous case, and

$$\lim_{N\to+\infty} \frac{1}{N}\sum_{n=0}^{N-1} |\mu(\phi^n A\cap B) - \mu(A)\cdot\mu(B)| = 0$$

[3] In contrast, if $\mu(M)=\infty$, A. B. Hajian [1] proved the following: let (M, μ, ϕ_t) be an ergodic system in which $\mu(M) = \infty$, A and B two measurable sets. For any $\varepsilon > 0$ there exist arbitrarily large t such that $\mu[\phi_t A\cap B] < \varepsilon$.

in the discrete case, for every pair of measurable sets A, B. R. V. Chacon (forthcoming paper) proved that if (M, μ, ϕ_t) is ergodic, then there exists a measurable change of the modulus of the velocity which makes the system weakly mixing.

V. A. Rohlin [1] has proposed the concept of n-fold mixing as a new invariant of the dynamical systems (see Halmos [1]): A dynamical system (M, μ, ϕ_t) is n-fold mixing, by definition, if

$$\lim_{\substack{\inf |t_i - t_j| \to +\infty \\ i \neq j}} \mu[\phi_{t_1}A_1 \cap \phi_{t_2}A_2 \cdots \cap \phi_{t_n}A_n] = \prod_{i=1}^{n} \mu(A_i)$$

for every n-tuple $A_1, \ldots, A_n$ of measurable sets. Mixing is a particular case ($n = 2$). Whether there exist mixing systems which are not n-fold mixing ($n > 2$) is an open question.

§9. Spectral Invariants

Let (M, μ, ϕ) be an abstract dynamical system. Let us denote the Hilbert space of the complex-valued functions defined on M with μ-summable square by $L_2(M, \mu)$. If $f, g \in L_2(M, \mu)$, we set

$$\langle f | g \rangle = \int_M f \cdot \bar{g} \cdot d\mu ,$$

where $\bar{z}$ is the complex conjugate of z, and

$$\|f\| = \sqrt{\langle f | f \rangle} .$$

DEFINITION 9.1

We set

$$Uf(x) = f(\phi(x)) , \tag{9.2}$$

where $f \in L_2(M, \mu)$. U *is a mapping that operates on functions, it is called the operator induced by* ϕ.

THEOREM 9.3 (KOOPMAN [1])

U is a unitary operator of $L_2(M, \mu)$.

Proof:

Since ϕ is measurable, U carries measurable, square summable functions into themselves. Hence, U maps $L_2(M, \mu)$ into $L_2(M, \mu)$.

(a) U is linear: For any $a, b \in \mathrm{C}$; $f, g \in L_2(M, \mu)$, we get:

$$U(af + bg) = (af + bg) \circ \phi = a(f \circ \phi) + b(g \circ \phi) = a \cdot Uf + b \cdot Ug \,.$$

(b) U is a bijection: Any $g \in L_2(M, \mu)$ can be written as some Uf. Explicitly $f(x) = g(\phi^{-1}x)$.

(c) U is isometric: Since ϕ is μ-measure preserving, setting $\phi y = x$, we get

$$\|Uf\|^2 = \int_M |f(\phi y)|^2 d\mu(y) = \int_M |f(\phi y)|^2 d\mu(\phi y)$$

$$= \int_M |f(x)|^2 d\mu(x) = \|f\|^2 \,. \qquad \text{(Q. E. D.)}$$

REMARK 9.4

In the continuous case (M, μ, ϕ_t) we obtain a continuous unitary one-parameter group U_t.

DEFINITION 9.5

It is clear that two dynamical systems (M, μ, ϕ) and (M', μ', ϕ'), which are isomorphic under f (see Definition 4.1), induce operators U and U' which are equivalent, that is, there exists an isomorphism F: $L_2(M, \mu) \to L_2(M', \mu')$ such that $U' = FUF^{-1}$, according to the following diagram (see Figure 9.6). *Thus, the invariants of U are certain invariants of the dynamical system (M, μ, ϕ). They are called the spectral invariants.* For instance, the spectrum of U is a spectral invariant.[4] A complete system

---

[4] In the continuous case U_t, it is the spectrum of the infinitesimal generator of U_t; or again, the spectrum associated with the resolution of the identity E for which, according to Stone's theorem: $U_t = \int_{-\infty}^{+\infty} e^{2\pi i \lambda t} dE(\lambda)$.

of spectral invariants is known (see Halmos [3]), the spectral measures and spectral multiplicities. Conversely, the equivalence of the operators U and U' (the corresponding dynamical systems are called spectrally equivalent) does not imply isomorphism (see Chapter 2, §12, Entropy; Appendix 15, Anzai Skew-products).

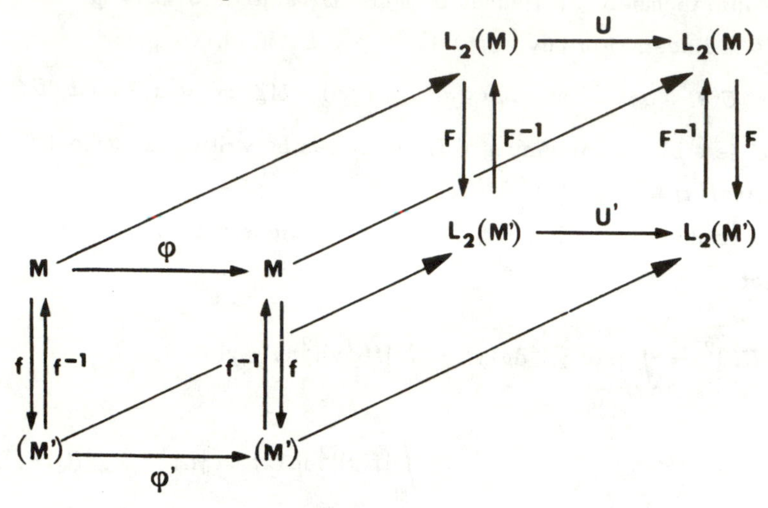

Figure 9.6

We now give some examples of ergodic properties that are reflected as spectral properties.

THEOREM 9.7 (ERGODICITY)

(M, μ, ϕ) is ergodic, if, and only if, 1 is a simple proper value of the induced operator U.

Proof:

If f lies in $L_2(M, \mu)$, then f is invariant if, and only if, $Uf = f$. Now ϕ is ergodic if, and only if, the invariant functions are all a.e. constant. Since the functions that are a.e. constant are scalar multiples of one another, ϕ is ergodic if, and only if, the subspace of solutions of $Uf = f$ has dimension 1. In the continuous case, the ergodicity of ϕ_t is equivalent to $\lambda = 0$ having multiplicity one in the spectrum of U_t.

THEOREM 9.8 (MIXING)

The dynamical system (M, μ, ϕ_t) *is mixing if, and only if,*

$$\lim_{T \to \infty} \langle U_t f \mid g \rangle = \langle f \mid 1 \rangle \cdot \langle 1 \mid g \rangle \tag{9.9}$$

for every $f, g \in L_2(M, \mu)$.

Proof:

If f and g are some characteristic functions, then (9.9) reduces to the very definition of mixing (8.2). The general case is derived easily by observing that the space of finite linear combinations of characteristic functions is dense in $L_2(M, \mu)$. In spectral terms, (M, μ, ϕ_t) is mixing if it is ergodic and the spectrum of U_t (except for $\lambda = 0$) is absolutely continuous with respect to the Lebesgue measure. The converse is false. We say that U_t has properly continuous spectrum if its only proper functions are constants. It can be proved (see Halmos [1]) that a dynamical system has properly continuous spectrum if, and only if, it is weakly mixing (see 8.9).

We turn to the case in which the spectrum of U_t is discrete.

EXAMPLE 9.10

Let M be the circle $\{z \mid z \in \mathbf{C}, |z| = 1\}$, μ its usual measure, ϕ the translation $\phi(z) = \theta \cdot z$, $\theta = e^{2\pi i \omega}$, $\omega \in \mathbf{R}$. Let us consider the function z^p, $p \in \mathbf{Z}$:

$$U z^p = (U z)^p = \theta^p \cdot z^p \ .$$

Hence, the z^p's are proper functions of U with corresponding proper values θ^p. The set $\{z^p \mid p \in \mathbf{Z}\}$, which is called the discrete spectrum of U, forms a complete orthonormal system of $L_2(M, \mu)$; whence the definition:

DEFINITION 9.11

A dynamical system (M, μ, ϕ) *has properly discrete spectrum if there is a basis of* $L_2(M, \mu)$, *each function of which is a proper function of the induced operator* U.

Let us turn back to (9.10). According to Theorem (9.7), the system is ergodic if, and only if, 1 is a simple proper value, that is, if, and only if, $p\omega \notin \mathbf{Z}$ when $p \neq 0$, which means ω is irrational. In otherwords, our system is ergodic if, and only if, the orbits are dense on M (see Example 7.8 and Appendix 1). Observe that ergodicity implies that all the proper values θ^p are distinct and simple. The system is not mixing; take $f = g = z^p$ in (9.8); we get:

$$< U^n z^p | z^p > = \theta^{pn} .$$

If $p \neq 0$, $\lim_{n=\infty} \theta^{pn}$ does not exist and (9.9) is not fulfilled. These results extend immediately to the n-dimensional torus and suggest the following theorem.

THEOREM 9.12

Let (M, μ, ϕ) *be an ergodic dynamical system,* U *the induced operator. Then:*

(a) *the absolute value of every proper function of* U *is constant a.e.;*

(b) *every proper value is simple;*

(c) *the set of all the proper values of* U *is a subgroup of the circle group* $\{z \,|\, z \in \mathbf{C},\ |z| = 1\}$;

(d) *if* (M, μ, ϕ) *is mixing, the only one proper value is* 1.

Proof:

Since U is unitary, every proper value λ has absolute value 1. It follows that if f is a corresponding proper function

$$f(\phi x) = \lambda f(x) \text{ a.e. implies } |f(\phi x)| = |f(x)| \text{ a.e.}$$

Hence, $f(x)$ is invariant under ϕ, and ergodicity implies that $|f|$ is constant a.e. (Corollary 7.6). In particular, $f \neq 0$ a.e.

Let h be another proper function with proper value λ. Since $f \neq 0$ a.e., h/f makes sense. We get:

$$U\left(\frac{h}{f}\right) = \frac{Uh}{Uf} = \frac{\lambda h}{\lambda f} = \frac{h}{f} ,$$

and h/f is an invariant function, so that h is a constant multiple of f. This proves (b).

If λ and μ are proper values of U, with corresponding proper functions f and g, we get

$$U\left(\frac{f}{g}\right) = \frac{Uf}{Ug} = \frac{\lambda f}{\mu g} = \lambda\mu^{-1}\cdot\left(\frac{f}{g}\right).$$

Hence f/g is a proper function of U with proper value $\lambda\mu^{-1}$. This proves (c).

Finally, if the system is mixing, take $f = g$ equal to a proper function with proper value in (9.9), we get:

$$\lim_{n=\infty} < U^n f \mid f > = < f \mid 1 > \cdot < 1 \mid f >,$$

that is

$$\lim_{n=\infty} \lambda^n = \text{constant}.$$

Hence $\lambda = 1$, and (d) is proved.

These properties of the discrete spectrum have been in some sense extended to the continuous part of the spectrum by Sinai [2], [3] (see, however, the recent paper of Katok and Stepin [1] for an example of a system whose maximal spectral measure does not dominate its convolution). The group of the proper values is obviously an invariant of the dynamical system. If the spectrum is discrete, this group forms a complete system of invariants. More precisely, we have:

DISCRETE SPECTRUM THEOREM 9.13 (VON NEUMANN, HALMOS)
(a) *Two ergodic dynamical systems with discrete spectrum are isomorphic if, and only if, the proper values of their induced operator coincide.*
(b) *Every countable subgroup of the circle group is the spectrum of an ergodic dynamical system with discrete spectrum.*

The proof will be found in Halmos [1]. It is based upon the construction of some compact abelian group (character group of the spectrum of given ergodic dynamical systems with properly discrete spectrum). Then, one proves the isomorphism of our given dynamical system with a translation of this abelian group.

To emphasize this result, we point out that the isomorphism problem is solved as far as the discrete spectrum case and abstract frame are concerned. In contrast, no characterization is known for the spectrum of a classical system. For instance, does there exist a classical system whose discrete spectrum is a prescribed subgroup of the circle group? Appendix 16 contains some information related to this question.

§10. Lebesgue Spectrum

Let us begin with an example.

EXAMPLE 10.1

We again consider Example (1.16): M is the torus $\{(x, y) \pmod 1\}$ with its usual measure; ϕ is the automorphism:

$$\phi(x, y) = (x + y, x + 2y) \pmod 1;$$

U is the induced operator. It is well known that the set

$$\mathbf{D} = \{e_{p,q}(x, y) = e^{2\pi i(px + qy)}, \quad p, q \in \mathbf{Z}\}$$

is an orthonormal basis of $L_2(M, \mu)$. The set $\mathbf{D}$ can be identified with the lattice $\mathbf{Z}^2 = \{(p, q)\} \subset \mathbf{R}^2$. Since $Ue_{p,q} = e_{p+q,\, p+2q}$, U induces an automorphism u on $\mathbf{D}$:

$$u: \begin{pmatrix} p \\ q \end{pmatrix} \to \begin{pmatrix} 1 & 1 \\ 1 & 2 \end{pmatrix} \begin{pmatrix} p \\ q \end{pmatrix} = \begin{pmatrix} p + q \\ p + 2q \end{pmatrix}.$$

Let us show that (0, 0) is the unique finite orbit of u. Assume that $(p, q) \in \mathbf{Z}^2$ has a finite orbit. This orbit is a bounded subset of R^2, invariant under the linear operator of $\mathbf{R}^2$,

$$\tilde{\phi} = \begin{pmatrix} 1 & 1 \\ 1 & 2 \end{pmatrix};$$

$\tilde{\phi}$ has two proper values λ_1, λ_2, $0 < \lambda_2 < 1 < \lambda_1$. Hence, $\tilde{\phi}$ is "dilating" in the proper direction corresponding to λ_1, and "contracting" in the proper direction corresponding to λ_2. This implies that the only invariant (under $\tilde{\phi}$) bounded subset of $\mathbf{R}^2$ is (0, 0). (Q. E. D.)

We conclude that $\mathbf{Z}^2 - \{0, 0\}$ splits into a set I of orbits of u, and each orbit is in an obvious one-to-one correspondence with $\mathbf{Z}$.

Let us go back to $\mathbf{D} = \{e_{p,q} \mid p, q \in \mathbf{Z}\}$. $\mathbf{D} - \{e_{0,0}\}$ splits into orbits of U: $C_1, C_2, \ldots, C_i, \ldots$; $i \in I$. If $f_{i,0}$ is some element of C_i, we may write

$$C_i = \{f_{i,n} \mid n \in \mathbf{Z}\},$$

where $f_{i,n} = U^n f_{i,0}$. To summarize, if H_i is the space spanned by the vectors of C_i, then $L_2(M, \mu)$ is the orthogonal sum of the H_i's and of the one-dimensional space of the constant functions. Each H_i is invariant under U and has an orthonormal basis $\{f_{i,n} \mid n \in \mathbf{Z}\}$ such that:

$$Uf_{i,n} = f_{i,n+1} .$$

Situations such as this occur often enough to deserve a definition.

DEFINITION 10.2

Let (M, μ, ϕ) be an abstract dynamical system, U the induced operator. (M, μ, ϕ) has Lebesgue spectrum L^I if there exists an orthonormal basis of $L_2(M, \mu)$ formed by the function 1 and functions $f_{i,j}$ $(i \in I,\ j \in Z)$ such that:

$$Uf_{i,j} = f_{i,j+1}, \text{ for every } i, j.$$

The cardinality of I can be easily proved to be uniquely determined and is called the *multiplicity of the Lebesgue spectrum.* If I is (countably) infinite, we shall speak of (countably) infinite Lebesgue spectrum. If I has only one element, the Lebesgue spectrum is called simple. An analogous definition holds in the continuous case. *Let U_t be the one-parameter group of induced operators of a dynamical flow (M, μ, ϕ_t). The flow is said to have Lebesgue spectrum L^I if every U_t $(t \neq 0)$ has Lebesgue spectrum L^I.*

REMARK 10.3

This terminology is derived from the following fact: Let

$$U_t = \int_{-\infty}^{\infty} e^{2\pi i t\lambda}\, dE(\lambda)$$

be the spectral resolution of U_t. It can be proved that (M, μ, ϕ_t) has

Lebesgue spectrum if, and only if, the measure $< E(\lambda)f|f >$ is absolutely continuous with respect to the Lebesgue measure, for every $f \in L_2(M, \mu)$ orthogonal to 1.

THEOREM 10.4

A dynamical system with Lebesgue spectrum is mixing.

Proof:

From Theorem (9.8) we need to prove that:

$$\lim_{n \to \infty} < U^n f | g > = < f|1 > \cdot < 1|g >$$

for every $f, g \in L_2(M, \mu)$. This is equivalent to:

$$\lim_{n = \infty} < U^n f | g > = 0$$

for every f, g orthogonal to 1. It is sufficient to prove this when f and g are basis vectors, for the general case follows by continuity and linearity. If $f = f_{i,j}$, $g = f_{k,r}$, then

$$< U^n f | g > = < f_{i,n+j} | f_{k,r} >$$

which is null for n large enough.

COROLLARY 10.5

The automorphism $\phi(x, y) = (x+y, x+2y) (\text{mod } 1)$ of the torus

$$M = \{(x, y) (\text{mod } 1)\}$$

(Example 1.16) has Lebesgue spectrum (Example 10.1). Then, it is mixing and ergodic (Corollary 8.4).

EXAMPLE 10.6

The Bernouilli schemes have countable Lebesgue spectrum. In particular they are spectrally equivalent.

Proof:

We prove it for $B(½, ½)$; the same statement holds for $B(p_1, \ldots, p_n)$ up to minor modifications. Let us recall (see Example 2.2) that $M = \mathbf{Z}_2^{\mathbf{Z}}$ is the space of the infinite bilateral sequences:

$$m = \ldots, m_{-1}, m_0, m_1, \ldots; \quad m_i \in \{0, 1\}.$$

The function 1 and the function

$$y_n(x) = \begin{cases} -1 & \text{if } x = 0 \\ +1 & \text{if } x = 1 \end{cases}$$

form an orthonormal basis of the space $L_2(\mathbf{Z}_2, \mu)$ associated to the n-th factor of M. From the product structure of M, we get an orthonormal basis of $L_2(M, \mu)$ which consists of the function 1 and all the finite products $y_{n_1} \cdot \ldots \cdot y_{n_k}$ of the y_n's with distinct indices $n_1, \ldots, n_k$. Now, let U be the induced operator of the shift ϕ. Call two elements of the above basis equivalent if some integer power of U carries one onto the other. The function 1 constitutes its own equivalence class; the other basis functions split into countably many equivalence classes. Each such equivalent class is in a one-to-one correspondence with $\mathbf{Z}$: the action of U on the class is to replace the element corresponding to $n \in \mathbf{Z}$ by the element corresponding to $n+1$.

To summarize, there exists an orthonormal basis of $L_2(M, \mu)$ consisting of the function 1 and of functions $f_{i,j}$ $(i = 1, 2, \ldots;\ j \in \mathbf{Z})$ such that

$$Uf_{i,j} = f_{i,\, j+1}$$

for every i, j. The number i is the number of the equivalence class, the number j is the number of its element which corresponds, as described above, to $j \in \mathbf{Z}$. Thus, $B(½, ½)$ has countable Lebesgue spectrum.

Let (M_1, μ_1, ϕ_1) and (M_2, μ_2, ϕ_2) be two Bernouilli schemes. There exist, from the above, orthonormal bases $\{1, f^1_{i,j}\}$ in $L_2(M_1, \mu_1)$ and $\{1, f^2_{i,j}\}$ in $L_2(M_2, \mu_2)$ such that:

$$U_1 f^1_{i,j} = f^1_{i,\, j+1}, \quad U_2 f^2_{i,j} = f^2_{i,\, j+1},$$

for every i, j. The isometry of $L_2(M_1, \mu_1)$ onto $L_2(M_2, \mu_2)$ defined by

$$1 \to 1, \quad f^1_{i,j} \to f^2_{i,j},$$

carries the spectral type of the first scheme into that of the other.

§11. *K*-Systems

In this section we define a class of abstract dynamical systems with strongly stochastic properties.

DEFINITION 11.1 [5]

An abstract dynamical system (M, μ, ϕ) *is called a K-system*[6] *if there exists a subalgebra* $\mathcal{A}$ *of the algebra of the measurable sets satisfying:*

(a) $$\mathcal{A} \subset \phi\mathcal{A},$$

(b) $$\bigcap_{n=-\infty}^{\infty} \phi^n \mathcal{A} = \hat{0},$$

where $\hat{0}$ *is the algebra of the sets of measure* 0 *or* 1,

(c) $$\overline{\bigvee_{n=-\infty}^{\infty} \phi^n \mathcal{A}} = \hat{1},$$

where ϕ, *by abuse of language, is the automorphism of* $\hat{1}$ *induced by* ϕ.

The above conditions become, in the continuous case:

(a′) $$\mathcal{A} \subset \phi_t\mathcal{A} \text{ for any } t \geq 0,$$

(b′) $$\bigcap_{t=-\infty}^{\infty} \phi_t \mathcal{A} = \hat{0},$$

(c′) $$\overline{\bigvee_{t=-\infty}^{\infty} \phi_t \mathcal{A}} = \hat{1}.$$

From the very definition, the isomorphic image of a *K*-system is a *K*-system.

EXAMPLE 11.2 BERNOUILLI SCHEMES (SEE 2.2)

The Bernouilli schemes are K-automorphisms.

Proof:

Let $B(p_1, \dots, p_n)$ be a Bernouilli scheme. The algebra $\hat{1}$ is generated

---

[5] See Appendix 17 for notations as $\subset, \wedge, \vee, \dots$. The standard notations (Rohlin) are: $\hat{1} = \mathfrak{M}$, $\hat{0} = \mathfrak{N}$.

[6] A. N. Kolmogorov [2] introduced this class under the name of quasi-regular systems.

by the:

$$A_i^j = \{m = \cdots m_{-1}, m_0, m_1, \ldots \mid m_i = j\}, \quad j \in Z_n, \quad i \in Z.$$

Let $\mathfrak{A}$ be the algebra generated by the A_i^j's, $i \leq 0$. We know that:

$$\phi(A_i^j) = A_{i+1}^j$$

where ϕ is the shift. Hence $\phi\mathfrak{A}$ is the algebra generated by the A_k^j's, $k \leq 1$, and

$$\mathfrak{A} \subset \phi\mathfrak{A},$$

proving the property (a). On the other hand, every generator A_r^j of $\hat{1}$ is a $\phi^q(A_i^j) = A_{i+q}^j$, $i \leq 0$, for $q = r - i$. Hence we get the property (b):

$$\overline{\bigvee_{n=-\infty}^{\infty} \phi^n\mathfrak{A}} = \hat{1}.$$

Let us now prove the property (c). Let $\mathfrak{B}$ be the subalgebra of $\hat{1}$, each element of which belongs to some subalgebra generated by a finite number of A_i^j. To every $A \in \mathfrak{B}$ there corresponds an $N \in Z$ such that $\mu(A \cap B) = \mu(A) \cdot \mu(B)$ for any $B \in \phi^{-n}\mathfrak{A}$, $n \geq N$ (exercise). Hence $\mu(A \cap B) = \mu(A) \cdot \mu(B)$ holds for every $B \in \bigcap_0^\infty \phi^{-n}\mathfrak{A}$. Since $\overline{\mathfrak{B}} = \hat{1}$, this relation still holds for any $A \in \hat{1}$ and $B \in \bigcap_0^\infty \phi^{-n}\mathfrak{A}$. Especially:

$$\mu(B) = \mu(B \cap B) = [\mu(B)]^2,$$

that is $\mu(B) = 0$ or 1, for every $B \in \bigcap_0^\infty \phi^{-n}\mathfrak{A}$. We conclude:

$$\bigcap_0^\infty \phi^{-n}\mathfrak{A} = \hat{0},$$

hence:

$$\hat{0} = \bigcap_{-\infty}^{\infty} \phi^n\mathfrak{A} \subset \bigcap_0^\infty \phi^{-n}\mathfrak{A}. \qquad \text{(Q. E. D.)}$$

COROLLARY 11.3

The Baker's Transformation is a K-system.

Proof:

This system is isomorphic to $B(\frac{1}{2}, \frac{1}{2})$ (see Appendix 7).

EXAMPLE 11.4

Chapter 3 will be devoted to a wide class of classical K-systems. This class contains the automorphisms of the tori, the geodesic flows on compact Riemannian manifolds with negative curvature, and the Boltzmann-Gibbs model of particles colliding elastically.

THEOREM 11.5

A K-system has a denumerably multiple Lebesgue spectrum. In particular, it is mixing and ergodic (Theorem 10.4). This theorem is due to Kolmogorov [2] for K-automorphisms and to Sinai [6] for K-flows. We sketch a proof for K-automorphisms, the complete proof will be found in Appendix 17.

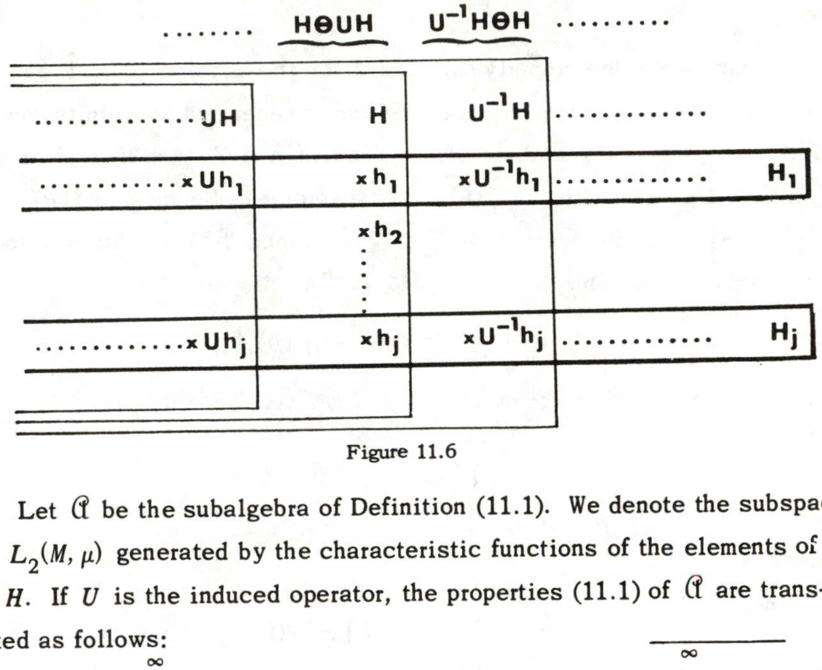

Figure 11.6

Let $\mathcal{A}$ be the subalgebra of Definition (11.1). We denote the subspace of $L_2(M, \mu)$ generated by the characteristic functions of the elements of $\mathcal{A}$ by H. If U is the induced operator, the properties (11.1) of $\mathcal{A}$ are translated as follows:

$$H_0 = \bigcap_{n=-\infty}^{\infty} U^n H \subset \cdots \subset UH \subset H \subset U^{-1}H \subset \cdots \subset \overline{\bigcup_{n=-\infty}^{\infty} U^n H}$$

$$= L_2(M, \mu),$$

where H_0 is the one-dimensional space of the constants.

Let us select an orthonormal basis $\{h_j\}$ on the orthocomplement $H \ominus UH$ of UH in H. H_j is the space spanned by the sequence $\ldots, U^{-1}h_j, h_j, Uh_j, \ldots$. The H_j's are invariant under U, and their orthogonal sum is $L_2(M, \mu) \ominus H_0$. Hence, if we set

$$e_{i,j} = U^j h_i, \quad i \in \mathbf{Z}^+, \quad j \in \mathbf{Z},$$

the $e_{i,j}$'s and the function 1 constitute an orthonormal basis of $L_2(M, \mu)$ such that:

$$Ue_{i,j} = e_{i,\, j+1}$$

for every i, j. We conclude that U has Lebesgue spectrum. The proof will be complete after it has been shown that the dimension of $H \ominus UH$ is infinite (see Appendix 17): U has a spectrum of infinite multiplicity.

§12. Entropy

This section is devoted to the definition and the study of a nonspectral invariant of dynamical systems introduced by A. N. Kolmogorov [4]. Throughout, $z(t)$ denotes the function on $[0, 1]$ defined by:

$$z(t) = \begin{cases} -t \operatorname{Log} t & \text{if } 0 < t \leq 1 \\ 0 & \text{if } t = 0 \end{cases}$$

where Log denotes the base-2 logarithm. We use the following properties of z: z is nonnegative, continuous, strictly concave

$$\left(z''(t) = -\frac{\operatorname{Log} e}{t} < 0\right),$$

and $z(t) = 0$ is equivalent to $t = 0$ or 1. Let $\alpha = \{A_i\}_{i \in I}$ be a *finite* measurable partition of M (see Appendix 18): I is finite and

$$\mu\left(M - \bigcup_{i \in I} A_i\right) = 0, \quad \mu(A_i \cap A_j) = 0$$

if $i \neq j$.

DEFINITION 12.1[7]

By definition, the entropy of the partition α is:

$$h(\alpha) = \sum_{i \in I} z(\mu(A_i)) \,.$$

EXAMPLE 12.2. PARTITION INTO N ELEMENTS OF EQUAL MEASURE

When $\mu(A_i) = 1/N$, then $h(\alpha) = \operatorname{Log} N$. Observe that if β is some other partition into N elements, then:

$$h(\beta) \leq \operatorname{Log} N$$

with equality if, and only if, each element has measure $1/N$. In fact, Jensen's inequality applied to the concave function z gives

$$h(\alpha) = \Sigma\, z(\mu(A_i)) = N\, \Sigma\, \frac{1}{N}\, z(\mu(A_i)) \leq N \cdot z\Big(\frac{1}{N}\, \Sigma\, \mu(A_i)\Big)$$

$$= N z\Big(\frac{1}{N}\Big) = \operatorname{Log} N \,.$$

Hence $h(\beta)$ is nothing but the weighted logarithm of the number of the elements of β.

Observe that if two partitions are equivalent, that is, if their elements coincide up to some sets of measure zero, their entropies are equal. In particular, the elements of measure zero can be removed. Finally, $h(\alpha) = 0$

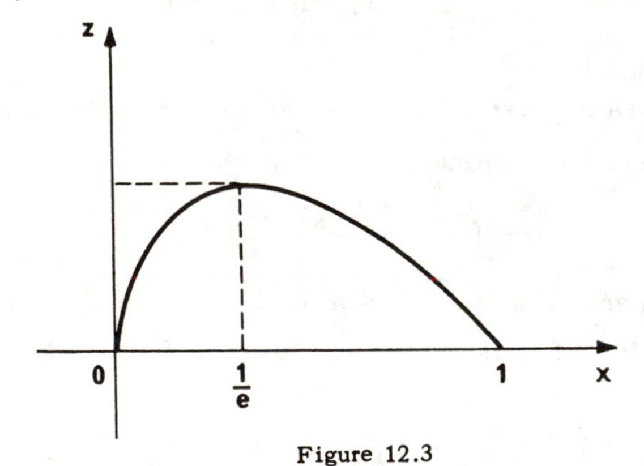

Figure 12.3

[7] For a probabilistic treatment see A. M. Yaglom — I. A. Yaglom [1].

means that $\alpha = \nu \pmod 0$ where ν is the trivial partition, the unique element of which is the space M.

DEFINITION 12.4. CONDITIONAL ENTROPY OF A PARTITION α WITH RESPECT TO A PARTITION β

Let $\alpha = \{A_i \mid i = 1,\dots,r\}$ and $\beta = \{B_j \mid j = 1,\dots,s\}$ be two finite measurable partitions (in the future, for short: partitions). *The entropy of* α *with respect to* β *is defined by:*

$$h(\alpha/\beta) = \sum_j \mu(B_j) \sum_i z(\mu(A_i/B_j)) ,$$

where

$$\mu(A_i/B_j) = \frac{\mu(A_i \cap B_j)}{\mu(B_j)}$$

is the conditional measure of A_i *relative to* B_j; α induces on each B_j a finite measurable partition α_{B_j}, the elements of which are $B_j \cap A_1, \dots, B_j \cap A_r$. After a suitable renormalization, B_j can be considered as a space of measure 1 on which α_{B_j} has entropy:

$$h(\alpha_{B_j}) = \sum_i z(\mu(A_i/B_j)) .$$

Hence, $h(\alpha/\beta)$ is the weighted sum of the $h(\alpha_{B_j})$'s:

$$h(\alpha/\beta) = \sum_j \mu(B_j) h(\alpha_{B_j}) .$$

The following theorem is proved in Appendix 18:

THEOREM 12.5

Let α, β, γ *be finite measurable partitions. Then:*

(12.6) $h(\alpha/\beta) \geq 0$ *with equality if, and only if* $\alpha \leq \beta \pmod 0$.

(12.7) $$h(\alpha \vee \beta/\gamma) = h(\alpha/\gamma) + h(\beta/\alpha \vee \gamma) .$$

(12.8) $$\alpha \leq \beta \pmod 0 \Longrightarrow h(\alpha/\gamma) \leq h(\beta/\gamma);$$

that is, conditional entropy is nondecreasing in its first argument.

(12.9) $$\beta \leq \gamma \pmod 0 \Longrightarrow h(\alpha/\gamma) \geq h(\alpha/\beta);$$

that is, conditional entropy is nonincreasing in its second argument.

(12.10) $$h(\alpha \vee \beta/\gamma) \leq h(\alpha/\gamma) + h(\beta/\gamma)\ ;$$

that is, conditional entropy is subadditive in its first argument.

Let ν be the trivial partition $\{M\}$ and let us take $\gamma = \nu$ in the above relations. Since $h(\alpha/\nu) = h(\alpha)$, we get:

(12.11) $$h(\alpha \vee \beta) = h(\alpha) + h(\beta/\alpha)\ ,$$

(12.12) $$\alpha \leq \beta \ (\mathrm{mod}\, 0) \Longrightarrow h(\alpha) \leq h(\beta)\ ,$$

(12.13) $$h(\alpha/\beta) \leq h(\alpha)\ ,$$

(12.14) $$h(\alpha \vee \beta) \leq h(\alpha) + h(\beta)\ .$$

Finally, if ϕ is an automorphism of the measure space (M, μ) and $\alpha = \{A_1, \dots, A_n\}$ is a partition, $\phi\alpha$ is a partition, namely:

$$\phi\alpha = \{\phi A_1, \dots, \phi A_n\}\ .$$

One verifies at once that:

(12.15) $$\phi(\alpha \vee \beta) = \phi\alpha \vee \phi\beta\ ,$$

(12.16) $$h(\alpha/\beta) = h(\phi\alpha/\phi\beta)\ .$$

DEFINITION 12.17. ENTROPY OF A PARTITION WITH RESPECT TO AN AUTOMORPHISM

Let (M, μ, ϕ) be a dynamical system, and α a finite measurable partition of M. By definition, *the entropy of α relative to ϕ is:*

$$h(\alpha, \phi) = \lim_{n=\infty} \frac{h(\alpha \vee \phi\alpha \vee \cdots \vee \phi^{n-1}\alpha)}{n}\ , \qquad n \in \mathbf{Z}^+\ .$$

Of course, we need to prove that this limit exists. Let n be a positive integer. We set:

$$h_n = h(\alpha \vee \phi\alpha \vee \cdots \vee \phi^{n-1}\alpha)\ ,$$

$$s_n = h_n - h_{n-1}\ .$$

LEMMA 12.18

$$s_n \geq 0\ .$$

Proof:

According to (12,12), $a \vee \phi a \vee \cdots \vee \phi^{n-1} a \leq a \vee \cdots \vee \phi^n a$ implies

$$h_{n-1} \leq h_n . \qquad \text{(Q. E. D.)}$$

LEMMA 12.19

$\{s_n\}$ is a nonincreasing sequence.

Proof:

According to (12.11):

$$s_n = h(a \vee \cdots \vee \phi^n a) - h(a \vee \cdots \vee \phi^{n-1} a) \tag{12.20}$$

$$= h(\phi^n a \,|\, a \vee \cdots \vee \phi^{n-1} a) .$$

Hence:

$$s_{n-1} = h(\phi^{n-1} a \,|\, a \vee \cdots \vee \phi^{n-2} a) ;$$

using (12.15) and (12.16) we get:

$$s_{n-1} = h(\phi^n a \,|\, \phi a \vee \cdots \vee \phi^{n-1} a) .$$

Since $\phi a \vee \cdots \vee \phi^{n-1} a \leq a \vee \cdots \vee \phi^{n-1} a$, (12.9) implies:

$$s_{n-1} \geq s_n .$$

THEOREM 12.21

$h(a, \phi)$ exists and is equal to $\lim_{n \to \infty} h(a \,|\, \phi^{-1} a \vee \cdots \vee \phi^{-n} a)$.

Proof:

s_n is a nonincreasing sequence of positive numbers: s_n has a limit s. Observe that $h_n = h(a) + s_1 + \cdots + s_n$, thus, Cesaro's mean convergence theorem implies:

$$\lim_{n = \infty} \frac{h_n}{n} = s .$$

Theorem (12.21) follows at once from (12.20) and the very definition of $h(a, \phi)$.

EXAMPLE 12.22. BERNOUILLI SCHEMES

Let $B(p_1, \ldots, p_k)$ be a Bernouilli scheme (see Example 2.2), ϕ the shift. We consider the finite partition a, the k elements of which are the

$$A_0^i = \{m \mid m_0 = i\}. \quad i = 1, \ldots, k.$$

We are going to prove that:

$$h(a, \phi) = -\sum_1^k p_i \operatorname{Log} p_i .$$

Since $\phi^n A_0^i = A_n^i = \{m \mid m_n = i\}$, the elements of $a \vee \phi a \vee \cdots \vee \phi^{n-1} a$ are the:

$$A_0^{i_0} \cap A_1^{i_1} \cap \cdots \cap A_{n-1}^{i_{n-1}} ,$$

whose measure is $p_{i_0} \cdots p_{i_n}$. Therefore:

$$h(a \vee \cdots \vee \phi^{n-1} a) = -\sum_{i_0, \ldots, i_{n-1}} p_{i_0} \cdots p_{i_{n-1}} \cdot \operatorname{Log}(p_{i_0} \cdots p_{i_{n-1}}) .$$

To carry out the summation over i_0, observe that

$$\sum_{i_1, \ldots, i_{n-1}} p_{i_1} \cdots p_{i_{n-1}} = 1 .$$

We get:

$$h(a \vee \cdots \vee \phi^{n-1} a) = -\sum_i p_i \operatorname{Log} p_i + h(a \vee \cdots \vee \phi^{n-2} a)$$

and by induction:

$$h(a \vee \cdots \vee \phi^{n-1} a) = n \left(-\sum_i p_i \operatorname{Log} p_i \right) ,$$

whence:

$$h(a, \phi) = -\sum_i p_i \operatorname{Log} p_i .$$

DEFINITION 12.23. ENTROPY OF AN AUTOMORPHISM [8]

The entropy $h(\phi)$ of an automorphism ϕ is:

$$h(\phi) = \sup h(a, \phi) ,$$

---

[8] Notion due to Kolmogorov [4]. See also Sinai [7], [8].

where the supremum extends over all finite measurable partitions α. It is clear that $h(\phi) \geq 0$.

THEOREM 12.24

$h(\phi)$ is an invariant of the dynamical system (M, μ, ϕ).

Proof:

Let (M', μ', ϕ') be a system isomorphic to (M, μ, ϕ). (see Definition 4.1). There exists an isomorphism $f\colon M \to M'$, that is $\phi' = f\phi f^{-1}$. If α is a partition of M, $f\alpha$ is a partition of M'. We get from (12.15) and (12.16):

$$h(f\alpha, \phi') = h(f\alpha, f\phi f^{-1}) = \lim_{n=\infty} \frac{h(f\alpha \vee \cdots \vee f\phi^{n-1} f^{-1} f\alpha)}{n}$$

$$= \lim_{n=\infty} \frac{h[f(\alpha \vee \cdots \vee \phi^{n-1}\alpha)]}{n} = \lim_{n=\infty} \frac{h(\alpha \vee \cdots \vee \phi^{n-1}\alpha)}{n} = h(\alpha, \phi) .$$

On the other hand, when α runs over all the partitions of M, $f\alpha$ runs over all the partitions of M'. We deduce at once:

$$\sup h(\alpha', \phi') = \sup h(\alpha, \phi) .$$

We now turn to a theorem which enables one, in many cases, to compute the entropy.

DEFINITION 12.25. GENERATOR WITH RESPECT TO AN AUTOMORPHISM

Let α be a finite measurable partition, $\mathfrak{M}(\alpha)$ the measure subalgebra generated by α:

α is called a generator with respect to ϕ if:

$$\bigvee_{n=-\infty}^{\infty} \phi^n \mathfrak{M}(\alpha) = \hat{1} .$$

KOLMOGOROV'S THEOREM[9] 12.26

If α is a generator relative to ϕ, then $h(\phi) = h(\alpha, \phi)$. The proof will

[9] See Kolmogorov [2], [4]; Sinai [7], [8].

be found in Appendix 19. Thus if ϕ possesses a generator, its entropy may be computed by the above formula. Let us give some examples:

EXAMPLE 12.27. BERNOUILLI SCHEMES

The entropy of the shift ϕ of $B(p_1, \ldots, p_k)$ is:

$$h(\phi) = -\sum_{i=1}^{k} p_i \operatorname{Log} p_i \ .$$

Proof:

Let a be the partition of 12.22, the elements of which are the

$$A_0^i = \{m \mid m_0 = i\}, \quad i = 1, \ldots, k \ .$$

Since $\phi^n A_0^i = A_n^i$, $n \in \mathbf{Z}$, the algebra $\bigvee_{-\infty}^{\infty} \phi^n \mathfrak{M}(a)$ contains all the generators of the algebra $\hat{1}$. Consequently a is a generator relative to ϕ. The above formula is a direct consequence of (12.22) and (12.26).

CONSEQUENCES 12.28

(1) Given an arbitrary nonnegative number a, there exists an abstract dynamical system, namely a Bernouilli scheme, the entropy of which is equal to a.

(2) We proved (Example 10.6) that the Bernouilli schemes are spectrally equivalent. But $B(1/2, 1/2)$ and $B(1/3, 1/3, 1/3)$ differ in their entropy and, according to Theorem (12.24) are not isomorphic. Hence, *there exist non-isomorphic abstract dynamical systems which are spectrally equivalent.*

It is conjectured that two K-systems are isomorphic if they possess the same entropy. Particular cases have been examined which indicate that this may be correct. We mention a result due to Meshalkin [1]: $B(p_1, \ldots)$ *and* $B(q_1, \ldots)$ *are isomorphic if they possess the same entropy and if the* p_i*'s,* q_i*'s are negative integral powers of one and the same positive integer* n. For instance, $B(1/2, 1/8, 1/8, 1/8, 1/8)$ and $B(1/4, 1/4, 1/4, 1/4)$ are isomorphic (here $n = 2$). Blum and Hanson [1] improved this result. In this direction we mention a theorem due to Sinai [9]: *Two K-systems with the same entropy are weakly-isomorphic, that is, each system is an homomorphic image of the other* (see Appendix 6).

EXAMPLE 12.29. AUTOMORPHISM OF THE TORUS

If ϕ is an ergodic automorphism of the torus $\{(x, y) \pmod 1\}$:

$$\phi(x, y) = (ax + by, cx + dy) \pmod 1, \qquad ad - bc = 1,$$

Sinai [7] proved that:

$$h(\phi) = \operatorname{Log} |\lambda_1| ,$$

where λ_1 is the proper value, whose modulus is greater than 1, of the matrix:

$$\begin{pmatrix} a & b \\ c & d \end{pmatrix} .$$

This result extends to the r-dimensional torus $T^r = \mathbf{R}^r/\mathbf{Z}^r$. *Let ϕ be an ergodic automorphism of T^r. If the matrix of ϕ has r distinct proper values $\lambda_1, \ldots, \lambda_r$, then:*

$$h(\phi) = \sum_{|\lambda_i| > 1} \operatorname{Log} |\lambda_i| .$$

(See Genis [1] and Abramov [1] for a correction of the proof.)

COROLLARY 12.30 (SEE EXAMPLE 4.4)

The dynamical systems defined on T^2 by:

$$\begin{pmatrix} 1 & 1 \\ 1 & 2 \end{pmatrix} \text{ and } \begin{pmatrix} 3 & 1 \\ 2 & 1 \end{pmatrix}$$

are nonisomorphic.

Those defined by:

$$\begin{pmatrix} 3 & 1 \\ 2 & 1 \end{pmatrix} \text{ and } \begin{pmatrix} 3 & 2 \\ 1 & 1 \end{pmatrix}$$

possess the same spectral type and the same entropy. Whether they are isomorphic is still an open problem. It is only known that they are weakly isomorphic (Sinai [9]).

THEOREM 12.31[10] ENTROPY OF K-SYSTEMS

The entropy of a K-automorphism is positive.

---

[10] Due to Kolmogorov [4].

Proof:

According to Definition (11.1) there exists a subalgebra $\mathcal{A}$ of $\hat{1}$ such that:

$$\hat{0} = \bigcap_{-\infty}^{\infty} \phi^n \mathcal{A} \subset \cdots \subset \mathcal{A} \subset \phi \mathcal{A} \subset \cdots \subset \bigcap_{-\infty}^{\infty} \phi^n \mathcal{A} = \hat{1} .$$

We first prove there exists a finite subalgebra $\mathcal{B} \subset \mathcal{A}$ *such that:*

$$\bigvee_{k \le n} \phi^k \mathcal{B} \supset \bigvee_{k \le n'} \phi^k \mathcal{B} \quad \textit{for } n > n'; \quad n, n' \in \mathbf{Z}. \tag{12.32}$$

Assume that for every finite subalgebra $\mathcal{B} \subset \mathcal{A}$ there exists $n, n' \in \mathbf{Z}$ such that $n > n'$ and

$$\bigvee_{k \le n} \phi^k \mathcal{B} = \bigvee_{k \le n'} \phi^k \mathcal{B} .$$

Since (M, μ) is a Lebesgue space, there exists an increasing sequence of finite subalgebras $\mathcal{B}_i$:

$$\mathcal{B}_1 \subset \mathcal{B}_2 \subset \cdots \subset \mathcal{A} ,$$

satisfying:

$$\overline{\bigvee \mathcal{B}_i} = \mathcal{A} .$$

From our assumption it follows that:

$$\phi^n \left(\bigvee_{k \le 0} \phi^k \mathcal{B}_i \right) = \bigvee_{k \le 0} \phi^k \mathcal{B}_i$$

for every i. As $\bigvee$ is associative and commutes with ϕ, we obtain:

$$\phi^n \mathcal{A} = \bigvee_{k \le n} \phi^k \mathcal{A} = \overline{\bigvee_i \phi^n \left(\bigvee_{k \le 0} \phi^k \mathcal{B}_i \right)} = \overline{\bigvee_i \left(\bigvee_{k \le 0} \phi^k \mathcal{B}_i \right)} = \mathcal{A} ,$$

which leads to a contradiction.

Now, let β be a finite partition generating the algebra $\mathcal{B}$ (see Appendix 18.5). From (12.21) we deduce:

$$h(\phi) \geq h(\beta, \phi) = \lim_{n=\infty} h(\beta \mid \phi^{-1}\beta \vee \cdots \vee \phi^{-n}\beta) .$$

The sequence $\phi^{-1}\beta$, $\phi^{-1}\beta \vee \phi^{-2}\beta, \ldots$, is nondecreasing, therefore, from (12.9) we get:

$$h(\phi) \geq h\left(\beta \mid \bigvee_{k \leq -1} \phi^k \beta\right).$$

Now, it is sufficient to show that $h(\beta \mid \bigvee_{k \leq -1} \phi^k \beta) > 0$. Assume that $h(\beta \mid \bigvee_{k \leq -1} \phi^k \beta) = 0$; this means (12.6):

$$\beta \leq \bigvee_{k \leq -1} \phi^k \beta \pmod 0 .$$

Consequently:

$$\bigvee_{k \leq 0} \phi^k \beta = \beta \vee \left(\bigvee_{k \leq -1} \phi^k \beta\right) = \bigvee_{k \leq -1} \phi^k \beta ,$$

that is

$$\bigvee_{k \leq 0} \phi^k \mathcal{B} = \bigvee_{k \leq -1} \phi^k \mathcal{B} ,$$

which contradicts (12.32). (Q. E. D.)

REMARK 12.33

Guirsanov [1] has constructed a nonclassical dynamical system with zero entropy and denumerably infinite Lebesgue spectrum. Hence it is not a K-system. Gourevitch [1] proved that the horocyclic flow on a compact surface with constant negative curvature has denumerably infinite Lebesgue spectrum and zero entropy. Hence, it is a classical system with Lebesgue spectrum, but it is not a K-flow.

REMARK 12.34

By definition, the entropy of a flow (M, μ, ϕ_t) is $h(\phi_1)$. If (M, μ, ϕ_t) is a K-flow (see Definition 11.1) then (M, μ, ϕ_1) is a K-automorphism. Consequently (Theorem 12.31) the entropy $h(\phi_1)$ of the K-flow is positive.

KOUCHNIRENKO'S THEOREM 12.35. ENTROPY OF CLASSICAL SYSTEMS[11]

Classical systems have finite entropy.

Proof:

Let (M, μ, ϕ) be a classical system. Since M is smooth, it carries some smooth Riemannian metric g. By a suitable conformal deformation $g \to e^{2\rho} \cdot g$, it is possible to assume that the volume element of g is precisely the measure element $d\mu$. The area of a submanifold will be the area in the sense of the metric g. By definition, a classical partition of M is a partition into a finite number of complexes with piecewise differentiable boundary. Since M is compact and smooth, such a partition always exists (Cairns [1]).

Let us begin with two obvious remarks:

(1) The classical partitions are dense in the sense of the entropy metric (see Appendix 19) in the set of the finite measurable partitions. Since $h(a, \phi)$ is continuous for the entropy metric (Appendix 19), we get:

$$h(\phi) = \sup_{a \text{ classical}} h(a, \phi) .$$

(2) Let σ be a $(\dim M-1)$-dimensional submanifold, the area of which is $S(\sigma)$. Since M is compact, there exists a constant λ (independent of σ) satisfying:

$$\frac{S(\phi\sigma)}{S(\sigma)} < \lambda . \tag{12.36}$$

Let a be a classical partition; $A_1, \ldots, A_n$ the elements of $a \vee \cdots \vee \phi^{k-1} a$. According to the isoperimetric inequality, there exists a constant C, which depends only on the manifold M, such that:

$$\mu(A_i)^{(N-1)/N} \leq C \cdot S(A_i) , \qquad N = \dim M .$$

Hence, we obtain:

---

[11] See Kouchnirenko [1], [2].

$$\sum_{i=1}^{n} [\mu(A_i)]^{(N-1)/N} \leq C \cdot \sum_{i=1}^{n} S(\partial A_i) \ .$$

If we denote by $S(\alpha)$ the sum of the areas of the boundaries of the elements of α (each boundary is counted twice), clearly:

$$\sum_{i=1}^{n} S(\partial A_i) = S(\alpha) + \cdots + S(\phi^k \alpha) \ ,$$

and then:

$$\sum_{i=1}^{n} [\mu(A_i)]^{(N-1)/N} \leq C[S(\alpha) + \cdots + S(\phi^{k-1}\alpha)] \ .$$

From (12.36) we deduce:

$$\sum_{i=1}^{n} [\mu(A_i)]^{(N-1)/N} \leq C \cdot S(\alpha)[1 + \cdots + \lambda^{k-1}] \leq C \cdot S(\alpha) \frac{\lambda^k}{\lambda - 1} \ ,$$

$$\text{Log} \ \Sigma \ [\mu(A_i)]^{(N-1)/N} \leq k \cdot \text{Log} \ \lambda + \text{constant.}$$

But Log t is concave, $\mu(A_i) \geq 0$, $\Sigma \ \mu(A_i) = 1$. Hence, Jensen's inequality applied to the left member gives:

$$\sum_{i=1}^{n} \mu(A_i) \ \text{Log} \ \mu(A_i)^{-1/N} \leq k \cdot \text{Log} \ \lambda + \text{constant,}$$

$$\frac{-\sum_{i=1}^{n} \mu(A_i) \ \text{Log} \ \mu(A_i)}{k} \leq N \cdot \text{Log} \ \lambda + \frac{\text{constant}}{k}$$

If $k \to +\infty$, we get:

$$h(\alpha, \phi) \leq N \cdot \text{Log} \ \lambda \ ,$$

and from remark 1:

$$h(\phi) \leq (\dim M) \cdot \text{Log} \ \lambda \ , \tag{12.38}$$

where the constant λ is given by relation (12.36). (Q. E. D)

COROLLARY 12.39

There exist abstract dynamical systems nonisomorphic to a classical one, for example, the infinite Bernouilli scheme:

$$B(1/2, 1/4, 1/16, 1/16, \ldots, \underbrace{1/2^{2n}, \ldots, 1/2^{2n}}_{2^{2^n-n-1}\text{ times}}, \ldots) .$$

COROLLARY 12.40

If an abstract system has discrete spectrum whose rank is finite, then the system has zero entropy.

Proof:

In fact the system is isomorphic to a compact, finite dimensional abelian group on which a translation ϕ acts. Hence $\lambda = 1$, and from (12.38): $h(\phi) = 0$.

A PROBLEM 12.41

Whether the entropy $h(\phi)$ of a classical system depends continuously on ϕ is an open question.

REMARK 12.42

Kouchnirenko's theorem is connected to recent results of M. Artin and B. Mazur [1]: Let M be a smooth compact manifold, then for a dense set of C^1-diffeomorphisms the number $N(n)$ of isolated fixed points of ϕ^n, $n = 1, 2, \ldots$, is exponentially bounded from above:

$$N(n) \leq C \cdot e^{\lambda n}, \qquad C = C(\phi), \quad \lambda = \lambda(\phi) .$$

REMARK 12.43

Recently, Kouchnirenko[12] introduced some new nontrivial invariants of abstract dynamical systems: A-entropies. Let A be a monotone sequence of integers

$$A: \; a_1 < a_2 < a_3 < \cdots .$$

---

[12] See his report at the Int. Math. Congr., Moscow, 1966.

Then the A-entropy of an automorphism ϕ with respect to a partition α is defined as:

$$h_A(\phi, \alpha) = \limsup_{n \to +\infty} \frac{h(\phi^{a_1}\alpha \vee \cdots \vee \phi^{a_n}\alpha)}{n} .$$

As in Definition (12.23), A-entropy is:

$$h_A(\phi) = \sup h_A(\phi, \alpha) .$$

One obtains the usual entropy if $A = \{0, 1, 2, \ldots\}$. The A-entropies can distinguish some systems with usual entropy 0. Let us give an example.

Let $A = \{2^n\}$. Then the A-entropy of the horocyclic flow (see Chapter 3) is $0 < h < \infty$. Consider the direct product of this flow onto itself. Its A-entropy is $0 < 2h < \infty$. Since $2h \neq h$, the product is not isomorphic to the horocyclic flow; however, they have both zero usual entropy and countable Lebesgue spectrum.

REMARK 12.44

Recently, Katok and Stepin [1] [13] introduced some new nontrivial invariants of abstract dynamical systems: periodical approximations speeds. Let (M, μ, ϕ) be an abstract dynamical system, let ξ_n be a partition of (M, μ) into sets C_n^i of measure $1/n$ $(i = 1, \ldots, n)$. An automorphism S_n of (M, μ) will be called cyclic with respect to the partition ξ_n if:

(a) $S_n \xi_n = \xi_n$;

(b) $(S_n)^n = E$ (identity), $(S_n)^k \neq E$ for $k < n$.

We shall say that ϕ admits approximation by cyclic transformations at the rate $0[f(q_n)]$ if, for an increasing sequence of natural integers q_n, there exists a sequence of partitions $\xi_{q_n} \to \hat{1}$ and a sequence of automorphisms S_{q_n}, cyclic with respect to ξ_{q_n}, such that

$$\sum_{i=1}^{q_n} \mu(\phi C_{q_n}^i \, \Delta \, S_{q_n} C_{q_n}^i) = 0[f(q_n)] .$$

Katok and Stepin have proved some important theorems which connect the

---

[13] See also their report at the Int. Math. Congr. Moscow, 1966.

notion of rate of approximation by periodic transformations with entropy and spectra. The importance of their results is related to the fact that, in many cases, it is possible to obtain some information on the approximation speeds of concrete systems, even if the explicit computation of spectra is impossible. Some of their theorems are the following:

(1) If the automorphism ϕ admits approximation by cyclic transformations at the rate $0(1/\ln^2 q_n)$, then $h(\phi) = 0$.

(2) An automorphism which admits approximation by cyclic transformations at the rate $0(1/q_n)$ is ergodic. Furthermore, strong convergence $U^{q_n} \Longrightarrow E$ occurs, where U is the unitary operator in $L^2(M, \mu)$ induced by ϕ. As a corollary, ϕ is not mixing and the maximal spectral type of U is singular.

One can find more theorems on approximations and their applications to the study of concrete dynamical systems, such as, for example, the mapping

$$\underbrace{\Delta_1 \quad \Delta_2 \quad \Delta_3} \longrightarrow \underbrace{\Delta_3 \quad \Delta_2 \quad \Delta_1}$$

or the flow on the torus

$$\frac{dx}{dt} = \frac{1}{F(x, y)}, \qquad \frac{dy}{dt} = \frac{\lambda}{F(x, y)}$$

in the papers of Katok and Stepin, in Doklady, in Founktzionalnyi analys i ego prilojeniia (1967), and in Uspehi (1967) Функциональный анализ и его приложения, Москва 1967.

General References for Chapter 2

Halmos, P. R., *Lectures on Ergodic Theory,* Chelsea (New York).

Halmos, P. R., *Entropy in Ergodic Theory: Lecture Notes,* University of Chicago (1959).

Hopf, E., *Ergodentheorie,* Springer, Berlin (1937).

Neumann, J. von, Zur operatoren Methode in der klassischen Mechanik, *Ann. Math.* 33 (1932) pp. 587–642.

Rohlin, V. A., New Progress in the Theory of Transformations with Invariant Measure, *Russian Math. Surveys* 15, No. 4 (1960) pp. 1–21.

Sinai, Ya., Probabilistic Ideas in Ergodic Theory, *Transl. Math. Soc.* Series 2, 31 (1963) pp. 62–84.

CHAPTER 3

UNSTABLE SYSTEMS

This chapter contains the study of classical systems with strongly stochastic properties, the so-called C-systems. [1]

The orbits of a C-system are highly unstable: two orbits with close initial data are exponentially divergent. This property turns out to imply the asymptotic independence of past and future: C-automorphisms are ergodic, mixing, have Lebesgue spectrum, have positive entropy, and, in general, are K-automorphisms. The set of the C-systems defined on a prescribed manifold M is an open set in the space of the classical systems defined on M. Consequently, every classical system, close enough to a C-system, is a C-system.

Geodesic flows on Riemannian manifolds with negative curvature are first examples of C-systems.

§13. C-Systems

EXAMPLE 13.1

Let us consider the torus $M = \{(x, y) \bmod 1\}$ and the diffeomorphism $\phi: M \to M$:

$$\phi: \begin{pmatrix} x \\ y \end{pmatrix} \to \begin{pmatrix} 1 & 1 \\ 1 & 2 \end{pmatrix} \begin{pmatrix} x \\ y \end{pmatrix} \quad (\text{mod } 1) .$$

The length $\|X\|$ of a tangent vector X is referred to the usual Riemannian metric $dx^2 + dy^2$ of M. Let $\phi^*: TM_m \to TM_{\phi m}$ be the differential of ϕ.

[1] Usually called U-systems in the English literature for, in Russian: "у-системы." Our terminology was introduced by Anosov; C is used because these systems satisfy a "condition C", in Russian "условие у."

In the chart (x, y), ϕ^* is the linear mapping:

$$\begin{pmatrix} 1 & 1 \\ 1 & 2 \end{pmatrix}$$

Hence, ϕ^* has two real proper values λ_1 and λ_2, $(0 < \lambda_2 < 1 < \lambda_1)$, with corresponding proper directions X and Y. The differential ϕ^* is dilating in the direction X and is contracting in the direction Y. To be precise, let X_m and Y_m be the subspaces of TM_m, respectively parallel to X and Y. Then:

$$\|\phi^*\xi\| \geq \lambda_1 \cdot \|\xi\| \text{ if } \xi \in X_m, \ (\lambda_1 > 1) ;$$
$$\|\phi^*\xi\| \leq \lambda_2 \cdot \|\xi\| \text{ if } \xi \in Y_m, \ (0 < \lambda_2 < 1) .$$

This is a characteristic example of the C-systems that we define next.

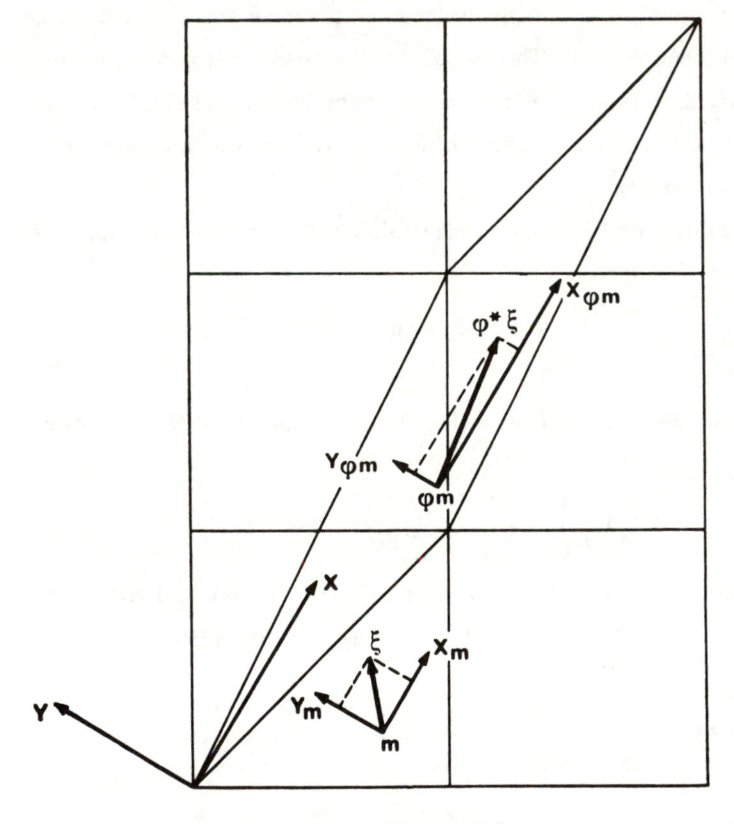

Figure 13.2

DEFINITION 13.3

Let ϕ be a C^2-differentiable diffeomorphism of a compact, connected, smooth manifold M. We denote the differential of ϕ by ϕ^*.

(M, ϕ) *is called a C-diffeomorphism if there exist two fields of tangent planes* X_m *and* Y_m *such that:*

(1) TM_m *falls into the direct sum of* X_m *and* Y_m*:*

$$TM_m = X_m \oplus Y_m, \quad \dim X_m = k \neq 0, \quad \dim Y_m = l \neq 0.$$

(2) *For every positive integer and for some Riemannian metric* g*:*

$$\|(\phi^n)^*\xi\| \geq a \cdot e^{\lambda n}\|\xi\|, \quad \|(\phi^{-n})^*\xi\| \leq b \cdot e^{-\lambda n}\|\xi\|, \quad \text{if } \xi \in X_m,$$

$$\|(\phi^n)^*\xi\| \leq b \cdot e^{-\lambda n}\|\xi\|, \quad \|(\phi^{-n})^*\xi\| \geq a \cdot e^{\lambda n}\|\xi\|, \quad \text{if } \xi \in Y_m.$$

The constants a, b, λ *are positive and independent for* n *and* ξ*, but* a *and* b *depend*[2] *on the metric* g. X_m is called the dilating space, and Y_m is the contracting space.

Example (13.1) is a C-diffeomorphism:

$$a = b = 1, \quad e^{\lambda} = \lambda_1, \quad e^{-\lambda} = \lambda_2 .$$

This definition extends to the continuous case ($t \in \mathbf{R}$): Let ϕ_t be a one-parameter group of C^2-differentiable diffeomorphisms of a compact, connected, smooth manifold M. (M, ϕ_t) *is called a C-flow if:*

(0) *the velocity vector* $\frac{d}{dt}\, \phi_t m|_{t=0}$ *does not vanish;*

(1) TM_m *splits into a direct sum:*

$$TM_m = X_m \oplus Y_m \oplus Z_m ,$$

where Z_m *is the one-dimensional space spanned by the velocity vector at* m*, and* $\dim X_m = k \neq 0$, $\dim Y_m = l \neq 0$;

---

[2] And hence for every metric: Let g_1 and g_2 be two Riemannian metrics on M. Due to the *compactness of M*, there exist two positive constants α, β such that: $\alpha\|\xi\|_2 \leq \|\xi\|_1 \leq \beta\|\xi\|_2$ for every $\xi \in TM$. Thus, if the inequalities 2 hold for g_1 with constants a and b, they still hold for g_2 with corresponding constants $(\alpha/\beta)a$ and $(\beta/\alpha)b$. This proves the independence of the definition from any metric g.

(2) *for any positive real number* t *and for some Riemannian metric* g:

$$\|(\phi_t)^*\xi\| \geq a\cdot e^{\lambda t}\|\xi\|,\ \|(\phi_{-t})^*\xi\| \leq b\cdot e^{-\lambda t}\|\xi\|,\ \text{if } \xi \in X_m;$$
$$\|(\phi_t)^*\xi\| \leq b\cdot e^{-\lambda t}\|\xi\|,\quad \|(\phi_{-t})^*\xi\| \geq a\cdot e^{\lambda t}\|\xi\|,\ \text{if } \xi \in Y_m.$$

The constants a, b, λ *are positive and independent for* t *and* ξ, *but* a *and* b *depend on the metric* g. Condition (0) means that the system has no equilibrium position. Condition (2) describes the behavior of the orbits. A C-diffeomorphism or a C-flow will be called, for short, a C-system.

REMARK 13.4

It is easy to show that:

(1) the subspaces X_m and Y_m are uniquely determined (they are, respectively, the "most dilating" and the "most contracting" subspaces of TM_m);

(2) $\dim X_m = k$ and $\dim Y_m = l$ do not depend on m (k is a continuous function of m, with integer values, on the connected space M);

(3) X_m and Y_m depend continuously on m.

Finally, observe that a C-system is not a classical system (see Definition 1.1) since we do not postulate the existence of an invariant measure. Now let us show how to construct certain C-flows from C-diffeomorphisms.

EXAMPLE 13.5 (SMALE)

The Space M.

Let $T^2 = \{(x, y) \bmod 1\}$ be the two-dimensional torus and $[0, 1] = \{u \mid 0 \leq u \leq 1\}$. We construct the cylinder $T^2 \times [0, 1]$, and after we identify $T^2 \times \{0\}$ and $T^2 \times \{1\}$ according to:

$$((x, y), 1) \equiv (\phi(x, y), 0) = ((x+y,\ x+2y), 0) \pmod 1,$$

where ϕ is the diffeomorphism (13.1):

$$\phi : \begin{pmatrix} x \\ y \end{pmatrix} \longrightarrow \begin{pmatrix} 1 & 1 \\ 1 & 2 \end{pmatrix}\begin{pmatrix} x \\ y \end{pmatrix} \pmod 1.$$

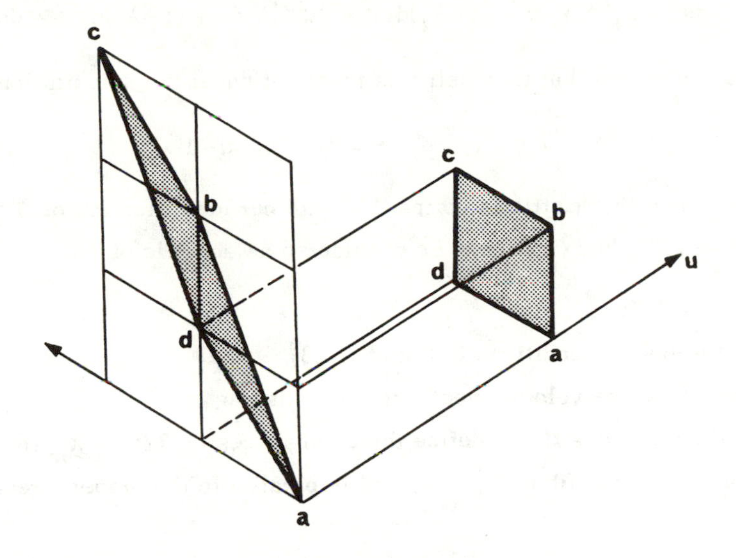

Figure 13.6

We obtain a compact manifold M. Let (x, y, u) be a point of M. The mapping $p: M \to S^1 = \{u \pmod 1\}$, $p(x, y, u) = u$ has rank 1 everywhere. Hence, M is a fibre bundle[3] with basis S^1 and fibre T^2.

The flow ϕ_t.

We define a flow ϕ_t by its infinitesimal generator:

$$\dot{x} = 0, \quad \dot{y} = 0, \quad \dot{u} = 1 . \tag{13.7}$$

An Auxiliary Riemannian Metric

Let λ_1 and λ_2, $(0 < \lambda_2 < 1 < \lambda_1)$, be the proper values of:

$$\begin{pmatrix} 1 & 1 \\ 1 & 2 \end{pmatrix} .$$

We define a Riemannian metric on $T^2 \times [0, 1]$ by:

---

[3] A differentiable fibre bundle (M, B, p) over B consists of the following: (i) a compact, connected, smooth $(n+q)$-dimensional manifold M; (ii) a smooth, n-dimensional manifold B called the base; (iii) a C^2-differentiable mapping $p: m \to B$ whose rank is n everywhere and called the projection. The $p^{-1}(b)$'s, $b \in B$, are called the fibres. They are q-dimensional manifolds diffeomorphic one to another.

$$(13.8) \qquad ds^2 = \lambda_1^{2u}[\lambda_1 dx + (1-\lambda_1)dy]^2 + \lambda_2^{2u}[\lambda_2 dx + (1-\lambda_2)dy]^2 + du^2 .$$

It is readily proved that this metric is invariant under the substitution;

$$x \to x + y, \quad y \to x + 2y, \quad u \to u - 1 .$$

In other words, this metric is compatible with our identification of $T^2 \times \{0\}$ and $T^2 \times \{1\}$. Thus (13.8) can be considered as a metric of M.

(M, ϕ_t) is a C-flow.

Take a look at conditions 0, 1, 2 of (13.3).

(0) From (13.3) the velocity vector is nonvanishing.

(1) If $m = (x, y, u) \in M$, we define three subspaces of TM_m: X_m (resp. Y_m) is tangent to the fibre $T^2 \times \{u\}$ and is parallel to the proper direction of ϕ:

$$\phi: \ T^2 \times \{u\} \to T^2 \times \{u\} ,$$

$$\phi: \begin{pmatrix} x \\ y \\ u \end{pmatrix} \to \begin{pmatrix} 1 & 1 & 0 \\ 1 & 2 & 0 \\ 0 & 0 & 1 \end{pmatrix} \begin{pmatrix} x \\ y \\ u \end{pmatrix} \ (\text{mod } 1)$$

with corresponding proper value λ_1 (resp. λ_2). Z_m is collinear to the velocity vector (13.7). Conditions (1) are fulfilled:

$$TM_m = X_m \oplus Y_m \oplus Z_m , \quad \dim X_m = \dim Y_m = 1.$$

(2) In the chart (x, y, u) the components of $\xi \in X_m$ are of the form:

$$(s, s(\lambda_1 - 1), 0), \qquad s \in \mathbf{R} .$$

On the other hand, according to (13.7), the matrix of ϕ_t^* reduces to the identity. We deduce from (13.8):

$$\|(\phi_t^*)\xi\|^2 = \lambda_1^{2(u+t)}[\lambda_1 s + (1-\lambda_1)(\lambda_1 - 1)s]^2 = \lambda_1^{2t} \cdot \|\xi\|^2 .$$

Hence, $\|(\phi_t^*)\xi\| = \lambda_1^t \|\xi\|$. This proves the first group of conditions (2) of (13.3):

$$a = b = 1, \qquad e^\lambda = \lambda_1 .$$

The second group is proved in the same way.

The field of two-planes $X_m \oplus Z_m$ (resp. $Y_m \oplus Z_m$) is clearly smooth and completely integrable: it defines a foliation[4] on M. Each sheet is the union of orbits of ϕ_t which are asymptotic to each other as $t \to -\infty$ (or $+\infty$; see Figure 13.9).

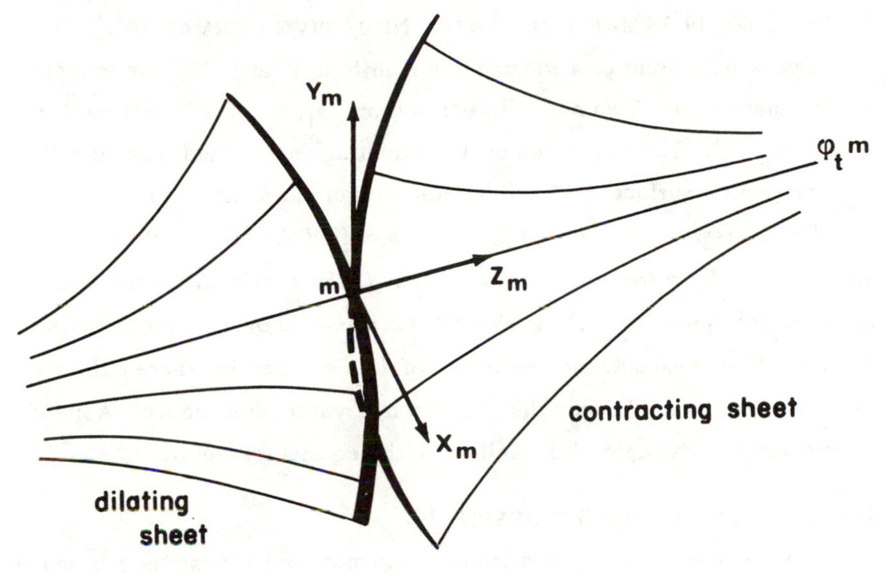

Figure 13.9

This property will be proved general for C-systems.

REMARK 13.10

The previous construction is quite general. Let (V, ϕ_0) be a C-diffeomorphism. We construct a compact manifold M by identification of $(v, 0)$ and $(\phi_0 v, 1)$ in the topological product $V \times [0, 1]$.

We define a C-flow over M by setting:

$$\phi_t(v, s) = (\phi_0^{[t+s]} v,\ t+s-[t+s]) ,$$

where $v \in V$, $s \in [0, 1]$, and $[a]$ means the integral part of a.

---

[4] By a foliated manifold M, we mean a completely integrable field of k-planes over M. The connected complete integral manifolds are called the sheets. They are k-dimensional submanifolds.

§14. Geodesic Flows on Compact Riemannian Manifolds with Negative Curvature

We turn next to an important example of C-system.

DEFINITION 14.1. MANIFOLDS WITH NEGATIVE CURVATURE[5]

Let v be a point of a Riemannian manifold V and TV_v the tangent vector space at v. Two noncollinear vectors e_1, e_2 of TV_v define a two-plane (e_1, e_2). The geodesics of V emanating from v and tangent to (e_1, e_2) generate a surface Σ. Σ is a Riemannian submanifold of V.

The Gaussian curvature of Σ *at* v *is called the sectional curvature* $\rho(e_1, e_2)$ *of* V *in the two-plane* (e_1, e_2). *If the sectional curvature is negative for every* (e_1, e_2), V *is called a manifold of negative curvature.* Then, if V is compact, the continuity of (e_1, e_2) implies there exists a constant $-k^2$ which bounds the sectional curvature from above. Appendix 20 provides an example of a manifold with negative curvature.

INSTABILITY OF THE GEODESICS 14.2

The geodesic flow ϕ_t of a Riemannian manifold V describes the movement of a material point on the frictionless manifold V without external forces [(see (1.7)]. If V has negative curvature the geodesics are very unstable: if $v, v_0 \epsilon T_1\tilde{V}$ [6], the distance $|\phi_t v, \phi_t v_0|$ increases exponentially with t. To be precise, we have the following result:

LOBATCHEWSKY-HADAMARD THEOREM[7] 14.3

Let V *be a compact, connected Riemannian manifold of negative curvature. Then, the geodesic flow on the unitary tangent bundle* $M = T_1 V$ *is a C-flow.*

Appendix 21 details the proof we sketch here (see Figure 14.4).

---

[5] For further information see S. Helgason[1], Chapter 1.

[6] $\tilde{V}$ = universal covering of V.

[7] This theorem is due to Lobatchewsky for surfaces of constant negative curvature. Hadamard [1] extended it to surfaces of arbitrary negative curvature.

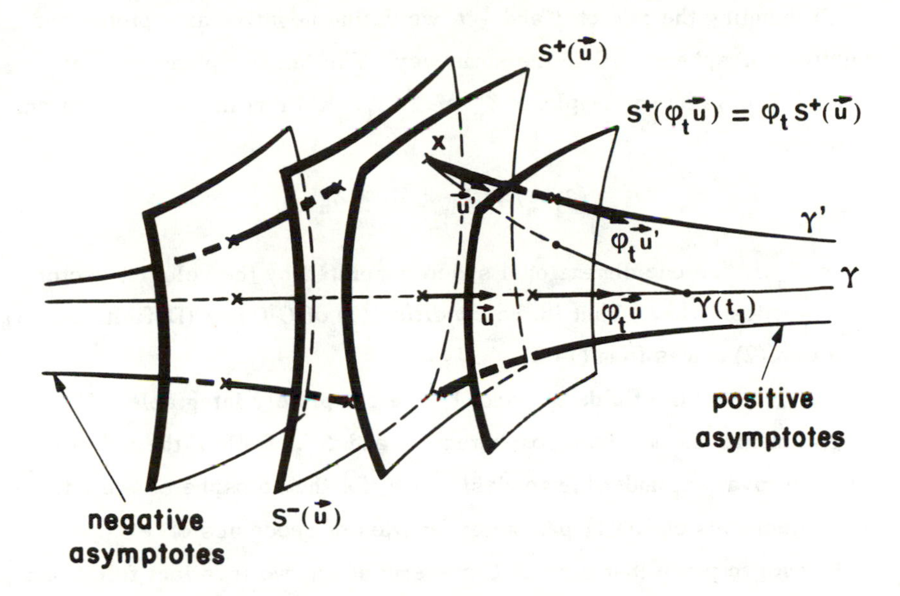

Figure 14.4

Let $\gamma(\mathbf{u}, t) = \gamma(t) = \gamma$ be a geodesic emanating from $\mathbf{u} \in T_1V$ and parametrized by arc length t. Let x be a point of V. There exists a geodesic γ_1 issuing from x and passing through $\gamma(t_1)$. As $t_1 \to +\infty$, γ_1 converges to a limit which is a geodesic $\gamma'(\mathbf{u}', t)$ emanating from $\mathbf{u}' \in TV_x$.

For a suitable choice of the origin of γ, it can be proved that:

$$\text{distance}\,(\gamma(t), \gamma'(t)) \leq b \cdot e^{-\lambda \cdot t}, \qquad t \geq 0, \tag{14.5}$$

where the constants b and λ are positive and independent for γ, γ', t. *Geodesics such as γ' are known as the positive asymptotes to γ.* They can be proved to be orthogonal trajectories to $(n-1)$-dimensional submanifolds ($n = \dim V$): the so-called *positive horospheres* S^+.

Let us denote by $S^+(\mathbf{u})$ the horosphere emanating from the origin of $\mathbf{u} \in T_1V$ and which is orthogonal to the positive asymptotes of $\gamma(\mathbf{u}, t)$. This horosphere can be interpreted as an $(n-1)$-dimensional submanifold of T_1V: $S^+(\mathbf{u})$ is the union of its orthogonal unitary vectors oriented as $\mathbf{u}$. The tangent plane at $\mathbf{u}$ of $S^+(\mathbf{u}) \subset T_1V$ is an $(n-1)$-plane $Y_{\mathbf{u}}$ of $T(T_1V_{\mathbf{u}})$.

Exchanging the role of t and $-t$, we define negative asymptotes and negative horospheres S^- in the same way. The tangent plane at u of $S^-(\mathrm{u}) \subset T_1V$ is an $(n-1)$-plane X_u of $T(T_1V_\mathrm{u})$. From the very definition, we have:

$$T(T_1V_\mathrm{u}) = X_\mathrm{u} \oplus Y_\mathrm{u} \oplus Z_\mathrm{u} ,$$

where Z_u is the one-dimensional space generated by the velocity-vector of the geodesic flow. That is the *condition* (1) *of C-flows* (Definition 13.3). Condition (2) comes from (14.5).

Observe that the fields X_u and Y_u are completely integrable. Their integral manifolds are the horospheres S^+ and S^-. Both of these foliations are invariant under the geodesic flow, for the horospheres are orthogonal trajectories of $(n-1)$-parameter families of geodesics of V.

We turn to prove that general C-systems admit two invariant foliations.

§15. The Two Foliations of a C-System

Let (M, ϕ) be a C-diffeomorphism; X_m (resp. Y_m) denotes the k-dimensional dilating space at $m \in M$ (resp. the l-dimensional contracting space). *A Riemannian metric on M is definitively selected.* Hence, X_m and Y_m are Euclidean subspaces of TM_m.

SINAI THEOREM[8] 15.1

Let (M, ϕ) *be a C-diffeomorphism, then:*

(1) *There exist two foliations* $\mathfrak{X}$ *and* $\mathfrak{Y}$ *that are invariant under* ϕ *and that are respectively tangent to the dilating field* X_m *and the contracting field* Y_m. *Hence, these fields are always integrable.*

(2) *Every diffeomorphism* $\phi': M \to M$, C^2*-close enough to* ϕ, *is a C-diffeomorphism. The dilating and contracting foliations* $\mathfrak{X}'$ *and* $\mathfrak{Y}'$ *of* ϕ' *depend continuously on* ϕ'.

---

[8] This was proved essentially in the paper by V. I. Arnold and Y. Sinai [6]; although their discussion was concerned with the particular case of small perturbations of automorphisms of a two-dimensional torus, the proof extend to the general case.

Appendix 22 completes the proof we sketch here.

CONSTRUCTION 15.2

The space K of the fields ρ of the tangent k-planes to M inherits a natural metric $|\rho_1 - \rho_2|$ which makes it into a complete metric space. Let ρ be such a field, and $\rho(m)$ the k-plane of TM_m. The diffeomorphism ϕ induces a mapping ϕ^{**}: $K \to K$:

$$\phi^{**}(m) = \phi^* \rho(\phi^{-1} m),$$

where ϕ^* is the differential of ϕ which maps a k-plane of TM_m onto a k-plane of $TM_{\phi(m)}$.

The dilating and contracting fields X and Y of ϕ are clearly fixed points of ϕ^{**}. It can be proved (see Appendix 22) that the axioms of C-systems imply that ϕ^{**} (or a positive integer power $(\phi^{**})^n$) is contracting in a neighborhood of the dilating field X:

$$|\phi^{**} \rho_1 - \phi^{**} \rho_2| \leq \theta |\rho_1 - \rho_2|, \quad 0 < \theta < 1, \tag{15.3}$$

for $|X - \rho_1| < \delta$, $|X - \rho_2| < \delta$, and δ small enough. Of course, (15.3) still holds for any diffeomorphism ϕ' C^2-close enough to ϕ, since ϕ'^* is C^1-close to ϕ^*. We deduce from the contracting mapping theorem that a mapping verifying (15.3) admits a fixed point. The fixed point of the mapping ϕ is X, but for ϕ' it is another field ρ'. Clearly:

$$\rho' = \lim_{n=\infty} (\phi'^{**})^n X,$$

$$\phi'^* \rho'(m) = \rho'(\phi' m),$$

and the field ρ' is dilating for ϕ'. A similar study of (M, ϕ^{-1}) leads to the contracting field of ϕ' that is close to Y. Thus ϕ' is a C.diffeomorphism.

THE INVARIANT FOLIATIONS 15.4

First assume that (M, ϕ) possesses two invariant foliations $\mathcal{X}$ and $\mathcal{Y}$, tangent, respectively, to X and Y. Then, the same property holds for (M, ϕ'). In fact, the invariant field ρ' of ϕ' is obtained as:

$$\rho' = \lim_{n=\infty} (\phi'^{**})^n X .$$

But $(\phi'^{**})^n X$ is the field of the k-planes tangent to the foliation $\phi'^n \mathcal{X}$. Thus, the limit field is integrable, that is tangent to a certain foliation $\mathcal{X}'$. This completes the proof of the part (2) of Theorem (15.1). As a matter of fact, the previous argument proves that there exists a foliation $\mathcal{X}$. Let us cover the compact manifolds M by a finite number of local charts (C_i, ψ_i); each C_i is a neighborhood of a point m_i and ψ_i is the mapping $\psi_i: C_i \to R^n$ $(n = \dim M)$.

Consider a foliation $\mathcal{X}_i^0$ of C_i such that $\psi_i(\mathcal{X}_i^0)$ be the foliation of $\psi_i(C_i)$; which consists in planes parallel to $\psi_i^*(X_{m_i})$. If the local charts are selected small enough, then the tangent plane $X_i^0(m)$ of $\mathcal{X}_i^0$ at $m \in C_i$ is close enough to the dilating plane X_m, for all m. Obviously there are several planes $X_i^0(m)$ passing through m if m belongs to several C_i.

Consider the foliations $\mathcal{X}_i^n = \phi^n \mathcal{X}_i^0$ of $\phi^n C_i$. These foliations cover M and (15.3) implies that their fields of tangent planes converge to X as $n \to +\infty$. We deduce readily that there exists a limit foliation $\mathcal{X}$ tangent to X. This concludes the proof.

REMARK 15.5

Each sheet of $\mathcal{X}$ is C^1-differentiable. But the foliation $\mathcal{X}$ is not required to be smooth: the normal derivative to the sheets may not exist. If $k = l = 1$, the field X is C^1-differentiable (see Arnold and Sinai [6]). It seems probable that $\mathcal{X}$ is not smooth in the general case. Anosov concocted an example where $\mathcal{X}$ and $\mathcal{Y}$ are not C^2-differentiable.

REMARK 15.6

The preceding proof extends to C-flows.

§16. Structural Stability of C-Systems

In this section we prove that C-systems are structurally stable.

DEFINITION 16.1. STRUCTURAL STABILITY

(A) *Diffeomorphisms*

Let M be a compact, smooth manifold and $\phi: M \to M$ a C^r-differentiable diffeomorphism.

By definition, ϕ is structurally stable if given a neighborhood $V(\mathrm{Id}_M)$ of the identity Id_M (in the C^0-topology[9]*) there exists a neighborhood $W(\phi)$ of ϕ (in the C^r-topology) such that if $\psi \in W(\phi)$, then there exists an homeomorphism $k \in V(\mathrm{Id}_M)$ making the following diagram commutative:*

$$\begin{array}{ccc} M & \xrightarrow{\ \phi\ } & M \\ k \downarrow \uparrow k^{-1} & & k^{-1} \uparrow \downarrow k \\ M & \xrightarrow{\ \psi\ } & M \end{array} ,$$

that is: $k \cdot \phi = \psi \cdot k$. In other words, k maps the orbits of $\{\phi^n \mid n \in \mathbf{Z}\}$ onto the orbits of $\{\psi^n \mid n \in \mathbf{Z}\}$.

(B) *Flows*

Let M be a compact smooth manifold and X a C^r-differentiable vector field which generates a flow ϕ_t:

$$X(m) = \frac{\mathrm{d}}{\mathrm{d}t} \phi_t m \Big|_{t=0} , \qquad m \in M .$$

By definition, ϕ_t is structurally stable if given a neighborhood $V(\mathrm{Id}_M)$ of the identity Id_M (C^0-topology) there exists a neighborhood $W(X)$ of X (C^r-topology) such that if $Y \in W(X)$, then there exists an homeomorphism $k \in V(\mathrm{Id}_M)$ that maps each orbit of X onto an orbit of Y. In the future we assume $r \leq 2$.

REMARK 16.2

One might restrict k to be a diffeomorphism rather than an homeomorphism. Then, consider in R^2 the system:

$$\dot{x} = y, \qquad \dot{y} = -x - Ky,$$

where K is a positive constant. The orbits are spirals focusing to the singular point $(0, 0)$ (see Figure 16.3). But, according to the Poincaré theory[10] of proper values, K is a continuous invariant of diffeomorphisms.

---

[9] Two mappings (or two vector fields) are C^r-close if their derivatives of order inferior to $r+1$ are close.

[10] See, for instance, K. Coddington and Levinson [1].

Consequently, two systems with distinct values of K would not be topologically conjugate and the focus $(0, 0)$ would not be structurally stable.

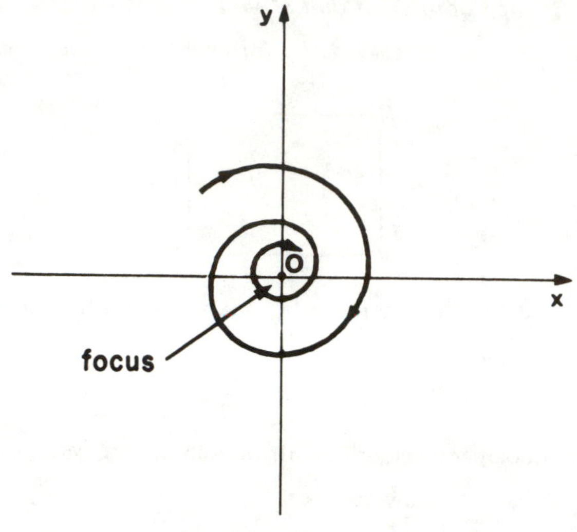

Figure 16.3

In the continuous case, one could be tempted to propose a definition quite similar to that of the discrete case: there exists an homeomorphism $k \in V(\mathrm{Id}_M)$ that makes the following diagram commutative for all t:

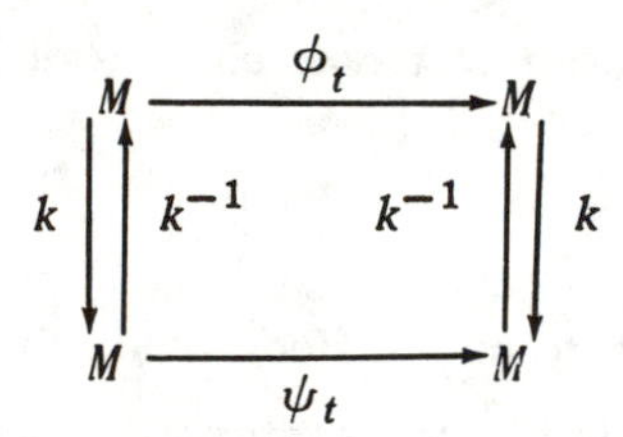

where ψ_t is the flow generated by Y. But, in such a definition, a limit cycle (see Figure 16.4) would not be structurally stable since its period is a continuous invariant (see footnote 10, p. 65).

Two problems arise naturally:

(1) What are the structurally stable phase-portraits?

(2) Given a manifold, are the structurally stable vector fields generic (or dense) in the space of the vector fields?

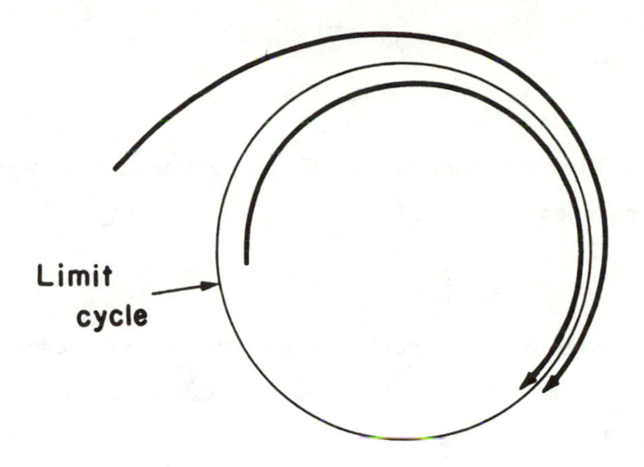

Figure 16.4

Andronov and Pontrjagin [1] gave an affirmative answer if M is the sphere S^2. The other two-dimensional manifolds were studied by Peixoto [1].

If the dimension is greater than two, the situation is involved. For instance, the system (13.1) is structurally stable but very complicated[11] (ergodic, the cycles are everywhere dense, and so on).

On the other hand, Smale [2] constructed an example (M, ϕ) such that: dim $M = 3$, every diffeomorphism close enough to ϕ is not structurally stable (see Appendix 24). This invalidates the genericity of structurally stable systems.

ANOSOV'S THEOREM[12] 16.5

C-Systems are structurally stable.

Sketch of the proof (see Appendix 25).

Let (M, ϕ) be a C-diffeomorphism, and $W(\phi)$ a neighborhood of ϕ (C^2-topology) in the space of the diffeomorphisms of M. We prove next that given $\phi' \in W(\phi)$, there exists a small and well-defined homeomorphism $k: M \to M$ such that:

---

[11] See S. Smale [1].

[12] See Anosov [1].

$$\phi' = k \cdot \phi \cdot k^{-1}$$

$$\sup_{m \in M} d[k(m), m] < \varepsilon ,$$

where d(,) is the Riemannian distance. We prove next that ε, which depends on ϕ', is bounded from above:

$$\sup_{\phi \in W(\phi)} \varepsilon < \varepsilon_1 .$$

Hence the structural stability holds good. In fact, we make correspond to ε_1 a neighborhood $W(\phi)$ such that given $\phi' \in W(\phi)$ there exists a homeomorphism $k: M \to M$ satisfying:

$$\phi' = k \cdot \phi \cdot k^{-1} ,$$

$$\sup_{m \in M} d[k(m), m] < \varepsilon_1 .$$

Let $\phi' \in W(\phi)$ be a diffeomorphism, C^2-close to ϕ. We already know that ϕ' is a C-diffeomorphism and ϕ and ϕ' have invariant dilating foliations $\mathfrak{X}$, $\mathfrak{X}'$ and invariant contracting foliations $\mathfrak{Y}$, $\mathfrak{Y}'$ (Sinai's theorem, §15.1). If there exists an ε-homeomorphism k: $M \to M$ such that $\phi' = k \cdot \phi \cdot k^{-1}$, setting $m' = km$, we have:

$$\text{(16.6)} \qquad d[\phi'^n(m'), \phi^n(m)] = d[k \cdot \phi^n(m), \phi^n(m)] < \varepsilon$$

for any $n \in \mathbf{Z}$.

From the very definition of C-systems we see that there exists at most one point m' satisfying (16.6). In fact, for any $\xi \neq 0$, we have:

$$\|(\phi'^*)^n \xi\| \to +\infty$$

in one or the other case $n \to +\infty$ (or $-\infty$). The uniqueness of the homeomorphism k being proved, if k exists we obtain it by making correspond to each point m the only point m' that satisfies (16.6) for all n. The problem is reduced to proving that such a point m' exists. The construction of m' is as follows.

The images $\phi'^n\beta$ $(n > 0)$ of each dilating sheet $\beta \in \mathfrak{X}'$, close to m, clearly have some points neighboring $\phi^n m$ (see Lemma A, Appendix 25 for the precise meaning). It can be proved (Lemma B, Appendix 25) that there exists, among these sheets, a unique sheet $\beta(m)$ such that its images $\phi'^n\beta(m)$ are still close to $\phi^n(m)$ for all $n < 0$.

The same device proves that there exists a unique contracting sheet $\delta(m) \in \mathcal{Y}'$ whose images $\phi'^n\delta(m)$ are close to $\phi^n(m)$ for all $n \in \mathbf{Z}$. The foliations $\mathcal{X}'$ and $\mathcal{Y}'$ are transverse; thus $\beta(m)$ and $\delta(m)$ intersect in a unique point m' in the neighborhood of m. It is easy to see that $\phi'(m') = (\phi(m))'$ and that m' varies continuously with m; thus the mapping $k: m \to m'$ is the desired homeomorphism.

C-FLOWS 16.7

Anosov's theorem extends to C-flow.

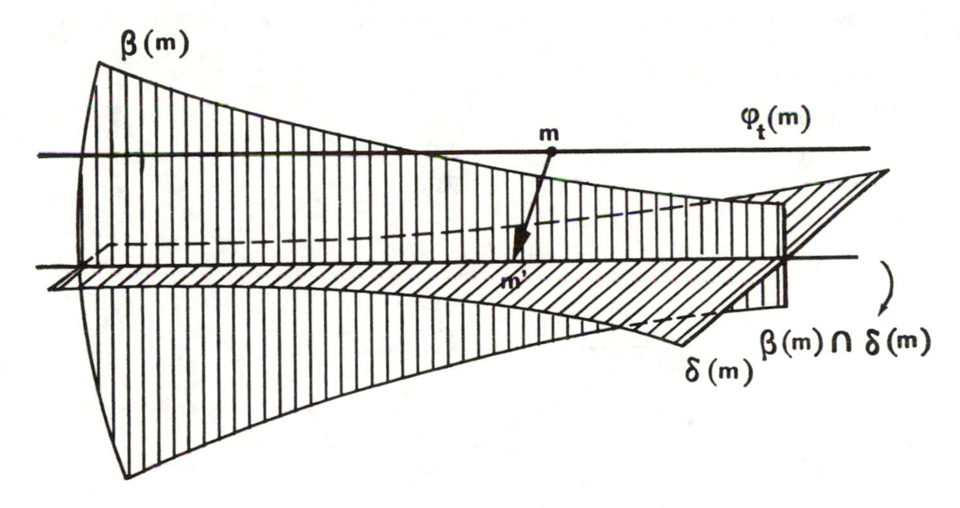

Figure 16.8

A similar construction (Figure 16.8) gives two sheets, $\beta(m)$ and $\delta(m)$. They are formed with orbits of ϕ_t' (respectively for $t \to +\infty$ and $t \to -\infty$) which are asymptotic to an orbit of ϕ_t' that is their intersection. As a geometric (nonparametrized) curve, this orbit is close to the orbit $\phi_t m$. The desired homeomorphism k is obtained by making correspond to m the nearest point of $\beta(m) \cap \delta(m)$ in the Riemannian sense.

§17. Ergodic Properties of C-Systems

Throughout this section we study C-systems that possess a positive and invariant measure μ. Thus, in contrast with the preceding sections, we are concerned with classical dynamical systems (see Definition 1.1).

Let us begin with ergodic properties of the automorphism (13.1): M is the torus $\{(x, y) \pmod 1\}$; the invariant measure is $d\mu = dx\,dy$; the automorphism is:

$$\phi: \begin{pmatrix} x \\ y \end{pmatrix} \to \begin{pmatrix} 1 & 1 \\ 1 & 2 \end{pmatrix} \begin{pmatrix} x \\ y \end{pmatrix} \pmod 1 .$$

THEOREM 17.1

(M, μ, ϕ) is a K-system

With a view to proving this theorem, we construct a subalgebra $\mathfrak{A}$ of $\hat{1}$.

THE SUBALGEBRA $\mathfrak{A}$. 17.2

The matrix:

$$\begin{pmatrix} 1 & 1 \\ 1 & 2 \end{pmatrix}$$

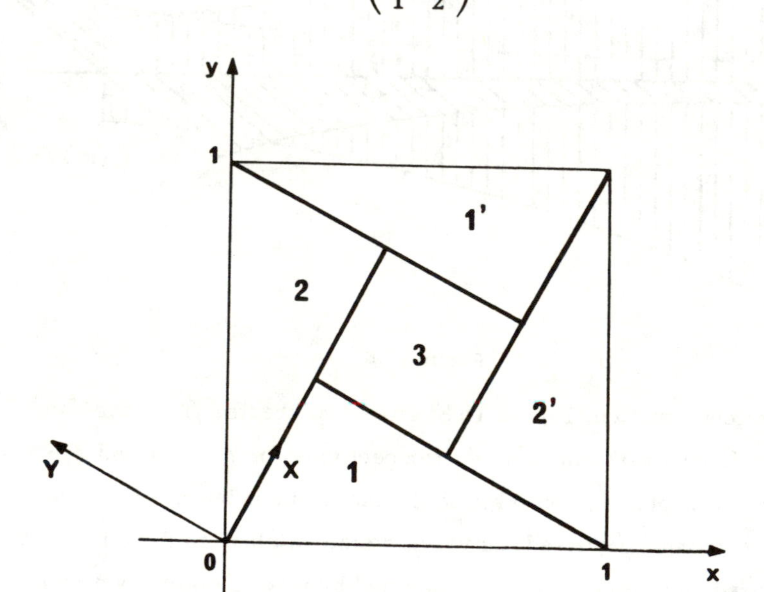

Figure 17.3

has two real proper values λ_1, λ_2 : $0 < \lambda_2 < 1 < \lambda_1$ with corresponding proper directions X and Y. We split the unit square of (x, y) into a square (3) and four triangles $(1, 2, 1', 2')$, with sides parallel to X and Y (see Figure 17.3). By pairwise identification of the opposite sides of the unit square, 1 and $1'$, 2 and $2'$, and 3 give rise to three disjoint parallelograms P_1, P_2, P_3 on the torus M. Again split P_1, P_2, and P_3 into parallelograms the sides of which are parallel to X and Y and whose lengths are inferior to a constant A. We get a partition β of M and we set:

$$a = \bigvee_{n=0}^{\infty} \phi^{-n}\beta .$$

The subalgebra $\mathcal{A}$ is the closure of the algebra $\mathfrak{M}(a)$ generated by a (see Appendix 18).

LEMMA 17.4

If the constant A is small enough, then:

(1) *The atoms of $\mathcal{A}$ are segments that are parallel to the contracting direction Y.*

(2) *Let l be a given positive number. The measure (in the sense of $dx\,dy$) of the union of those atoms whose length is inferior to l is bounded from above by $C \cdot l$, where C is an absolute constant. In particular, almost every atom of $\mathcal{A}$ is a segment and does not reduce to a point.*

Proof:

We argue in the covering plane (x, y). An element of $\beta \vee \phi^{-1}\beta$ is of the form:

$$B_1 \cap \phi^{-1} B_2 ; \quad B_1, B_2 \in \beta .$$

$\phi^{-1}B_2$ is a parallelogram with sides parallel to X and Y and whose lengths are, respectively, inferior to $\lambda_1^{-1} \cdot A$ and $\lambda_2^{-1} \cdot A$. On M, the intersections $B_1 \cap \phi^{-1}B_2$ results from the intersection on $(x, y) = \tilde{M}$ of $\phi^{-1}B_2$ with six parallelograms B_1, TB_1, $T^{-1}B_1$, tB_1, tTB_1, $tT^{-1}B_1$, that are deduced from $B_1 \subset [0, 1] \times [0, 1]$ by the translations $T = (1, 0)$, $t = (-1, 0)$. The projections of these six parallelograms into

$$X \subset (x, y) = \tilde{M}$$

are six segments, the lengths of which are inferior to A and whose mutual distances are greater than $K - A$, where:

$$K = \inf |k \cos \theta - k' \sin \theta| ,$$

as $|k| + |k'| \neq 0$; $k = -1, 0, 1$; $k' = -2, -1, 0, 1, 2$; and θ is the angle of X with $0y$. Consequently, if $2A$ *is inferior to* K, $\phi^{-1}B_2$ intersects at most one of those six parallelograms. Thus, on M, $B_1 \cap \phi^{-1}B_2$ is a connected set: it is a parallelogram with sides parallel to X and Y, the lengths of which are, respectively, inferior to $\lambda_1^{-1} \cdot A$ and A.

Exchanging β and $\beta \vee \phi^{-1}\beta$, the above argument applies for

$$\mathrm{Max}(\lambda_1^{-1}A, A) \leq A .$$

Thus, the elements of:

$$(\beta \vee \phi^{-1}\beta) \vee \phi^{-1}(\beta \vee \phi^{-1}\beta) = \beta \vee \phi^{-1}\beta \vee \phi^{-2}\beta$$

are parallelograms with sides parallel to X and Y and the lengths of which are, respectively, inferior to $\lambda_1^{-2} A$ and A.

By induction on n, we see that the elements of $\beta \vee \phi^{-1}\beta \vee \cdots \vee \phi^{-n}\beta$ are parallelograms with sides parallel to X and Y and the lengths of which are, respectively, inferior to $\lambda_1^{-n}A$ and A. As $n \to +\infty$, we obtain the first part of Lemma (17.4), $\lambda_1 > 1$. Let L be the sum of the lengths of the sides of the elements of β that are parallel to X. The analogous sum for $\phi^p\beta$, $p \in \mathbf{Z}$, is $\lambda_1^p \cdot L$. Thus, the analogous sum for α is bounded from above by:

$$L + \lambda_1^{-1} \cdot L + \cdots = \frac{L}{1-\lambda_1^{-1}} \overset{\text{def}}{=} C.$$

Some of the elements of α have a side parallel to Y with length inferior to l. Consider the set of these elements and denote by m the measure of their union. From the preceding formula we deduce at once that $m \leq C \cdot l$; this proves part (2).

THE CONTRACTING FOLIATION 17.5

Consider a nonnull constant vector field on M whose vectors are parallel to the contracting direction Y. *The integral lines are called the contracting sheets* (see §15). *This foliation is called ergodic if any union of sheets, with positive measure, coincides with M up to a set of measure zero.*

LEMMA 17.6

The contracting foliation is invariant under ϕ and ergodic.

Proof:

The invariance follows from the invariance of Y. The contracting sheets are the orbits of a translation group of M. To prove ergodicity it is sufficient to prove that each sheet is everywhere dense (Appendix 11) or, equivalently, is not closed (Jacobi theorem; Appendix 1). Assume there exists a closed contracting sheet F with length f. The invariance of the foliation implies that $\phi^n F$ is again a closed sheet with length $\lambda_2^n \cdot f$. Since $0 < \lambda_2 < 1$, we have $\lambda_2^n \cdot f \to 0$ as $n \to +\infty$. But this contradicts the obvious fact that the length of any sheet has to be greater than 1.

Proof of Theorem 17.1:

According to the very definition of $\mathcal{A}$: $\mathcal{A} \subset \phi\mathcal{A}$ this is the condition (a) of Definition (11.1). Let us prove the condition (c):

$$\overline{\bigvee_{n=0}^{\infty} \phi^n \mathcal{A}} = \hat{1} .$$

It is sufficient to prove that the atoms of

$$\bigvee_{n=0}^{\infty} \phi^n a$$

have arbitrarily small length. According to Lemma (17.4) the atoms of $a \vee \cdots \vee \phi^n a$ are of the form:

$$A_0 \cap \phi A_1 \cap \cdots \cap \phi^n A_n ; \quad A_0, \ldots, A_n \in a ,$$

where the A_i's are segments parallel to Y and with length inferior to A. Since A is small enough, then $A_0 \cap \phi A_1 \cap \cdots \cap \phi^n A_n$ is a connected segment (see argument of Lemma (17.4)). The length of this segment is inferior to $\lambda_2^n \cdot A$, for ϕ is contracting in the direction Y $(0 < \lambda_2 < 1)$. As $n \to +\infty$, we have $\lambda_2^n A \to 0$; this proves the condition (c).

We now turn to the condition (b). Let H be an element of positive measure of

$$\bigcap_{-\infty}^{\infty} \phi^n \mathfrak{A} .$$

H is the union of atoms of $\mathfrak{A}$, the union of atoms of $\phi^{-1}\mathfrak{A}$, $\phi^{-2}\mathfrak{A}$, and so on. Let ε be an arbitrary positive number inferior to 1. Take a number l such that:

$$C \cdot l < \varepsilon \cdot \mu(H),$$

where C is the constant of Lemma (17.4). Up to a set E of measure inferior to $\varepsilon \cdot \mu(H)$, the atoms of $\mathfrak{A}$ have a side parallel to Y whose length is greater than l. Since ϕ^{-1} is measure-preserving and dilating in the direction Y with ratio λ_2^{-1}, one concludes that, up to a set $\phi^{-n}E$ of measure inferior to $\varepsilon \cdot \mu(H)$, the atoms of $\phi^{-n}\mathfrak{A}$ have a side parallel to Y whose length is greater than $\lambda_2^{-n} \cdot l$. As $n \to +\infty$, we obtain—up to a set of measure inferior to $\varepsilon \cdot \mu(H)$ —H is the union of contracting sheets. From the arbitrariness of ε and from the ergodicity of the contracting foliation (Lemma (17.6)), one deduces at once that $\mu(H) = 1$. Whence:

$$\bigcap_{-\infty}^{\infty} \phi^n \mathfrak{A} = (\emptyset, M) \ (\text{mod } 0) \qquad \text{(Q. E. D.)}$$

The above arguments extend to general C-diffeomorphisms. The delicate point consists in constructing a partition similar to β. A theorem due to Anosov [2] overcomes this difficulty. First we need a definition.

Consider a foliation of an n-dimensional Riemannian manifold M into p-dimensional sheets. Let us take any small element of an $(n-p)$-

dimensional manifold Π transversal to the sheets. Assume that near Π there is another such manifold Π' and that each sheet passing through $m \in \Pi$ intersects Π' in a point m' near m in the induced Riemannian metric of the sheet. Consider the map $f\colon \Pi \to \Pi'$ taking m to m'.

DEFINITION 17.7

If for any such pair of neighboring manifolds the map f has a continuous generalized Jacobian, and if for small deformations of Π' this Jacobian varies continuously, then the foliation is called absolutely continuous.

For both of the two foliations of a C-system (see §15), Anosov proved:

THEOREM 17.8

The foliations $\mathfrak{X}$ and $\mathfrak{Y}$ are absolutely continuous.

We cannot give here the proof of Theorem (17.8), which is too long. We only announce the final results obtained with its help. The statement of Theorem (17.1) extends and leads to the following theorems (Anosov [2]).

THEOREM 17.9

Every C-system (C-diffeomorphism or C-flow) is ergodic.

THEOREM 17.10 , (SEE SINAI [11])

Every C-diffeomorphism is a K-system.

For C-flows the situation is more complicated. The C-flow ϕ_t of Example (15.5) has nonconstant proper functions. Consequently, ϕ_t does not have Lebesgue spectrum (see §10) and cannot be a K-flow (Theorem 11.5). The following theorem proves that, in some sense, Example (15.5) is the only one exception.

THEOREM 17. 11

Let ϕ_t be a C-flow on a compact, n-dimensional manifold M, then:

(1) either ϕ_t is a K-system; or

(2) ϕ_t has nonconstant proper functions.

In case (2), such a proper function is continuous and there exists a compact, $(n-1)$-dimensional submanifold V of M and a C-diffeomorphism

$\phi: V \to V$ such that ϕ_t is the one obtained from ϕ by means of the construction described in (13.10), *up to a change of the time scale t ($t \to C \cdot t$, C = constant).*

COROLLARY 17.12

The geodesic flow on the unitary tangent bundle of a Riemannian manifold V of negative curvature is a K-system.

Proof:

According to Theorem (14.3), the geodesic flow ϕ_t is a C-flow. Hence, from Theorem (17.9), ϕ_t is ergodic.

On the other hand, by passing to the double covering, we may assume V orientable. If V is two-dimensional, it is not diffeomorphic to the torus T^2 since the Gauss-Bonnet formula implies that the Euler-Poincaré characteristic is negative. Consequently, according to Corollary (A 16.10) of Appendix 16, the only continuous proper functions of ϕ_t are constants. The second case of Theorem (17.11) cannot occur and this theorem proves that ϕ_t is a K-flow. [13]

Thus, ϕ_t has positive entropy (see (12.31)), has denumerably multiple Lebesgue spectrum [14] (see (11.5)), is mixing [15] (see (10.4)), and ergodic [16] (see (8.4)).

§18. Boltzmann-Gibbs Conjecture

The methods and ideas of the preceding sections are applicable to certain problems of Classical Mechanics, for example to the Boltzmann-Gibbs model of a gas. This model consists of hard spheres contained in a parallelepipedic box with rigid walls. Collisions between spheres, or spheres

---

[13] This result is due to Sinai [10] for compact surfaces of negative curvature.

[14] This result is due to Gelfand and Fomin [2] for arbitrary dimension and constant negative curvature.

[15] This result is due to E. Hopf [1] in the case of constant negative curvature.

[16] This result is due to Hedlund [1] and E. Hopf [1] for surfaces of negative curvature and manifolds of constant negative curvature.

and wall, are supposed to be perfectly elastic. Sinai has proved ergodicity[17] of this system on each manifold T = constant $\neq 0$.

Ergodicity derives from the collisions. As the simplest model let us consider the motion of two perfectly elastic circular particles on the surface of a two-dimensional torus having a Euclidean metric. For simplicity we consider one of the particles as fixed. The second particle (which can now be regarded as a point) moves on a "torus billiard table" (Figure 18.1), being reflected from the fixed circumference according to the law "the angle of incidence α is equal to the angle of reflection β."

Let us, at the same time, consider an elliptic billiard table (Figure 18.2). The ellipse can be regarded as an oblate ellipsoid on which the

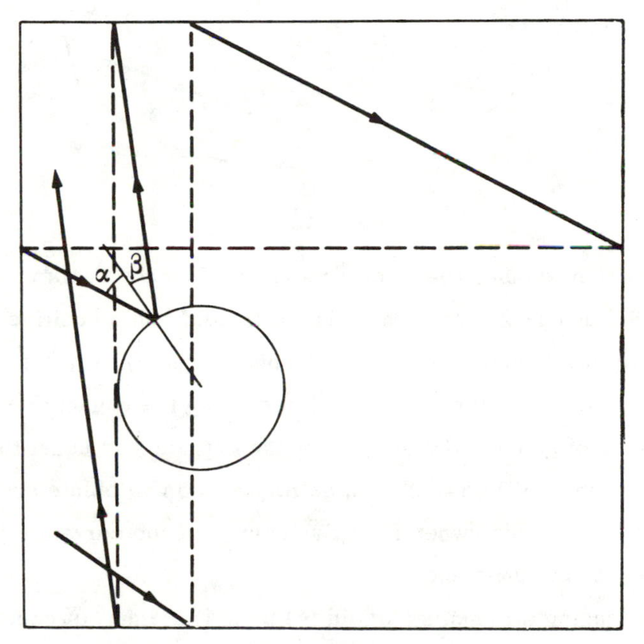

Figure 18.1

---

[17] This result was conjectured as the "quasi-ergodic hypothesis." At the beginning of Ergodic Theory this conjecture occasioned sharp discussions. But the problem was, perhaps, overvalued: statistical mechanics deals with asymptotic behavior as $N \to +\infty$ (N = number of particles) and not as $t \to +\infty$ for fixed N.

point moves along a geodesic, passing at each reflection from one side to the other. In the same way a torus billiard table can be regarded as a two-sided torus with a hole on which the point moves along a geodesic. But if the two-sided ellipse is an oblate ellipsoid, the two-sided torus with a hole will be an oblate surface of genus 2. Thus, motion on our torus billiard table is a limiting case of geodesic flow on a surface of genus 2.

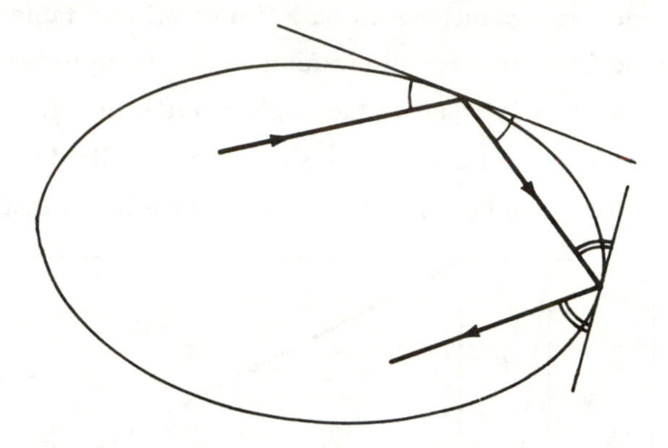

Figure 18.2

We now turn to our billiard tables and consider the curvature to which Figures 18.1 and 18.2 correspond. The ellipsoid has a positive curvature whose integral is equal to 4π (Gauss-Bonnet formula). On flattening the ellipsoid, all the curvature is accumulated along the boundary of an ellipse. For a surface of genus 2 the integral of the curvature is equal to -4π. Thus, a two-sided torus billiard table can be regarded as an oblate surface with negative curvature everywhere: on flattening, all the curvature is accumulated along the circumference.

The preceding arguments (Arnold [4] p. 184) are not, of course, a proof of ergodicity, not even in the simplest cases we considered. Nevertheless, using methods and notions dealing with C-systems (asymptotic orbits, transverse foliations), Sinai [4], [5] succeeded in proving that the Boltzmann-Gibbs model is ergodic on each submanifold T = constant $\neq 0$ and, even more, is a K-system.

The proofs of these results require hundreds of pages and contain a theory of generalized *C*-systems with discontinuous foliations. We only mention that the general case reduces to a "billiard table" problem in the configuration space: the collision-hypersurface of the configuration space plays the part of our circumference.

General References for Chapter 3

Anosov, D. V., Geodesic Flows on Compact Riemannian Manifolds of Negative Curvature, *Trudy Instituta Steklova* 90 (1967).

Hopf, E., Statistik der geodätischen Linien in Mannigfaltigkeiten negativer Krümmung, *Ber. Verh. Sächs. Akad. Wiss.*, Leipzig 91 (1939) pp. 261–304.

Sinai, Ya, Dynamical Systems with Countably Multiple Lebesgue Spectra II. *Isvestia Math. Nauk.* 30 No. 1 (1966) pp. 15–68.

CHAPTER 4

STABLE SYSTEMS

Many dynamical systems are known, the orbits of which, with remarkable stability, fail to fill up the "energy level" $H = Ct$ ergodically and remain (to the end) in their particular corner of phase-space. The systems that are close to an "integrable" system, and the systems to which the Theory of Perturbations of Celestial Mechanics applies, belong to this class. Let us mention, for instance, the three-body problem, the fast rotations of a heavy rigid body, the motion of a free point in a geodesic on a convex surface, and the adiabatic invariants.

Only in recent times, beginning with the work of C. L. Siegel [1] (1942) and A. N. Kolmogorov [5] (1954), has some progress been made in studying these systems. This chapter reports on the present status of this problem. We refer to V. Arnold [4], [5] for further details.

Let us begin with an example.

§19. The Swing and the Corresponding Canonical Mapping

EQUATIONS OF THE MOTION 19.1

Pendulum equations are:

$$\dot{q} = p, \quad \dot{p} = -\omega^2 \sin q, \tag{19.2}$$

where ω is the "proper frequency" that depends on the length of the pendulum.

A swing is a pendulum whose length l periodically varies (Figure 19.3). Its equations can be written as

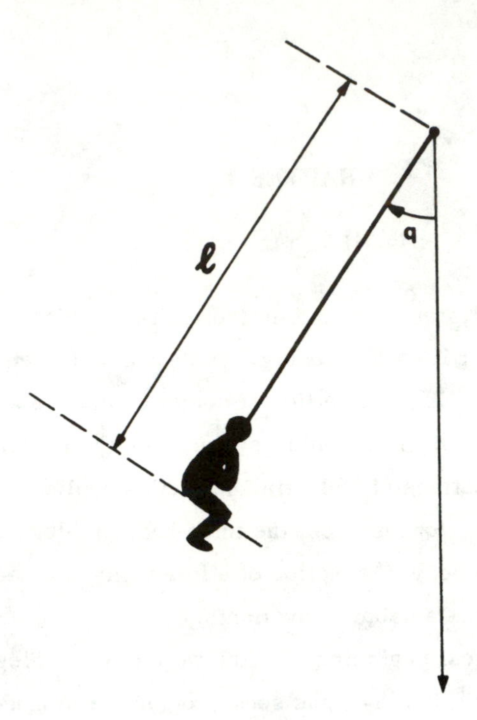

Figure 19.3

$$\dot{q} = p, \quad \dot{p} = -\omega^2(t) \sin q, \tag{19.4}$$

where $\omega(t+\tau) \equiv \omega(t)$. Appendix 5 studies (19.2) by using the phase-plane. Equations (19.4) contain the time variable t explicitly. Thus, we deal with the study of a vector field in the three-dimensional space p, q, t (see Figure 19.5).

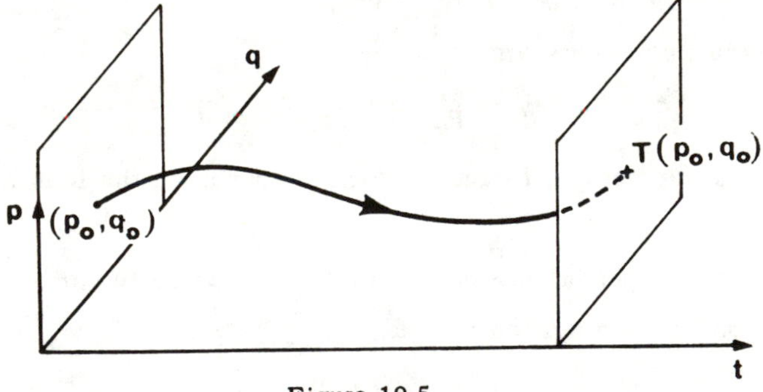

Figure 19.5

THE MAPPING T 19.6

The initial data $p(0) = p_0$, $q(0) = q_0$ define an orbit $p = p(t)$, $q = q(t)$. The time periodicity of Equations (19.4) allows one to identify the surfaces $t = 0$ and $t = \tau$. Thus, (19.4) can be regarded as equations in the space $\mathbf{R}^1 \times S^1 \times S^1 = \{p, q \bmod 2\pi, t \bmod \tau\}$. This defines a mapping T of the surface Σ $(t=0)$ into itself:

$$T(p_0, q_0) = (p(\tau), q(\tau)) .$$

Clearly:

$$p(n\tau), q(n\tau) = T^n(p_0, q_0),$$

and the study of $p(t)$, $q(t)$ as $t \to +\infty$ reduces to the study of the iterates T^n, $n \in \mathbf{Z}$. The mapping T is canonical, for the Equations (19.4) are canonical. Therefore, T preserves the area $dp \wedge dq$.

The equilibrium positions $p = 0$, $q = k \cdot \pi$ $(k=0, 1)$ are solutions of (19.4). Thus, the points $p = 0$, $q = k\pi$ are fixed points of T.

THE INTEGRABLE CASE 19.7

Let us begin with the "integrable case" $\omega =$ constant, to get acquainted with the mapping T. In this case, the system is conservative; thus, the energy is invariant. In other words, on the surface Σ, the curves (Figure A5.1, Appendix 5):

$$\Gamma: \ \tfrac{1}{2}p^2 - \omega^2 \cos q = h$$

are invariant under T.

Let us consider the part of Σ that is interior to the separatrix $(h < \omega^2)$. We use "action-angle variables I, ϕ" to study T. It can be proved (Appendix 26) that there exists a canonical transformation $p, q \to I, \phi$ such that the equation $I = Ct$ defines an invariant curve $\Gamma = \Gamma_I$. The coordinate ϕ (mod 2π) is the angular coordinate of Γ_I and, in coordinates I, ϕ, the mapping T becomes:

$$T: \ I, \phi \to I, \phi + \lambda(I) .$$

This means that the curve Γ_I rotates through an angle $\lambda(I)$ that is constant along each curve Γ_I (if ϕ is selected as parameter), but depends on the curve. It is readily seen that $\lim \lambda(I) = 0$ as the curve Γ_I converges to the separatrix $\frac{1}{2}p^2 - \omega^2 \cos q = \omega^2$, and that[1] $\lim \lambda(I) = \omega\tau$ as Γ_I converges to $(0, 0)$. Hence, some curves Γ_I are rotated through an angle $\lambda(I)$ commensurate with 2π, and others through an angle incommensurate with 2π. Let us consider the iterates of T. If $x = (p, q)$ belongs to a curve Γ such that $\lambda = 2\pi \frac{m}{n}$, then: $T^n x = x$. Thus, each point of Γ is a fixed point of T^n, and the orbit of x consists in a finite number of points (Figure 19.8). If the angle $\lambda(I)$ that corresponds to Γ_I is incommensurate

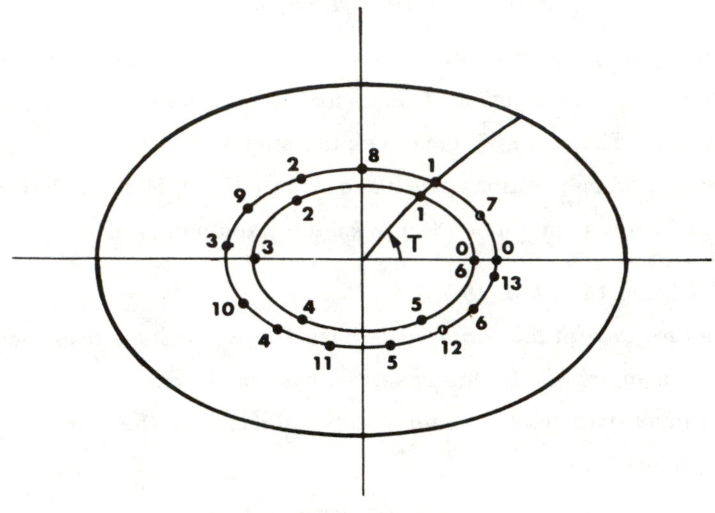

Figure 19.8

with 2π, then the points $T^n x$ are everywhere dense on Γ_I (Appendix 1).

Finally, observe that the equilibrium position $p = q = 0$ is stable: if

$$|x_0| = \sqrt{p_0^2 + q_0^2}$$

is small enough, then $|T^n x_0|$ remains small for any $n \in \mathbf{Z}$.

---

[1] The linear part of T at zero $(p, q \ll 1)$: $\dot{q} = p$, $\dot{p} = -\omega^2 q$ is a rotation with frequency ω (see Appendix 5). This gives rise to a rotation with angle $\omega\tau$ after time τ.

NONINTEGRABLE CASES 19.9

Suppose that ω is a nonconstant periodic function. We assume additionally that $\omega(t)$ is close to a constant ω_0, for instance:

$$\omega^2 = \omega_\varepsilon^2(t) = \omega_0^2(1 + \varepsilon \cos \nu t), \qquad 0 < \varepsilon << 1, \quad \nu = 2\pi/\tau .$$

The mapping T_ε, which corresponds to ω_ε, is close to the above mapping T. This mapping T_ε preserves the area $dp \wedge dq$ and the point $(0, 0)$, but neither the energy nor the curves Γ_I are preserved. The main purpose of the Theory of Perturbations is to study the behavior of the iterates T_ε^n as $n \to +\infty$ and $\varepsilon << 1$. There are two ways to consider the problem:

(1) $\varepsilon << 1, \ -\infty < n < \infty$ *(Theory of Stability);*

(2) $\varepsilon << 1, \ |n| < \varepsilon^{-k}$ *(Asymptotic Theory of the k-th approximation).*

The main result of the Theory of Stability is due to Kolmogorov (see §21). In our example, it reads as follows:

THEOREM 19.10

If ε is small enough, then the mapping T_ε has invariant analytical curves Γ_ε that are close to the invariant curves Γ of T. Besides, for ε small enough, these curves Γ_ε fill up the domain interior to the separatrices ($\frac{1}{2}p^2 - \omega_0^2 \cos q \leq \omega_0^2$), up to a set whose Lebesgue measure is small with ε . Roughly speaking, for ε small enough, the invariant curves Γ_I, whose angles $\lambda(I)$ are "incommensurate enough" with 2π, do not collapse but are only slightly deformed. The images $T_\varepsilon^n x$ of $x \in \Gamma_\varepsilon$ are all contained in Γ_ε .

To an invariant curve Γ_ε of T_ε corresponds a torus[2] of the space p, q, t ($q \bmod 2\pi$, $t \bmod \tau$) that is invariant under (19.4). These tori divide the space p, q, t (see Figure 19.11). Therefore a trajectory of (19.4) beginning in a gap between two invariant tori cannot pass out of this gap. Thus, Theorem (19.10) allows us to reach conclusions regarding the stability of motion.

---

[2] That is a conditionally periodic motion of the swing.

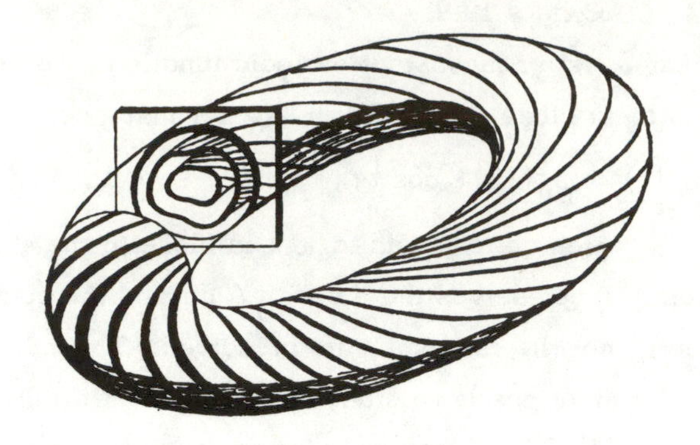

Figure 19.11

Theorem (19.10) is a direct corollary of Theorem (21.11) whose proof can be found in Appendix 34.

§20. Fixed Points and Periodic Motions

Let us study the fixed points of the mapping T_ε and its iterates, to understand better the structure of the gaps between two invariant curves Γ_ε. These points correspond to the periodic motions of the swing.

THE ELLIPTIC AND HYPERBOLIC POINTS 20.1

In the neighborhood of a fixed point, a mapping reduces to its linear part, that is, its differential, up to terms of second order. The differential of a canonical mapping is a linear canonical mapping. Linear canonical mappings are studied in Appendix 27. If this linear mapping is hyperbolic (resp. hyperbolic with reflection, resp. elliptic), then the fixed point is called hyperbolic (resp. hyperbolic with reflection, resp. elliptic).

Hyperbolic fixed points are easily proved to be unstable, not only for the linear mapping, but also for the nonlinear mapping (Hadamard). The problem of stability of elliptic points is known as the Birkhoff problem. Points of elliptic type are, in general, stable in the two-dimensional case (see Appendix 28).

STABILITY OF THE LOWEST POSITION OF EQUILIBRIUM 20.2

Now, consider the fixed point $p = q = 0$ of the mapping T_ε of §19. If $\varepsilon = 0$, then the mapping $T_\varepsilon = T$ is elliptic at the origin; in fact, the differential of T is an elliptic rotation with angle $\lambda = \omega\tau = 2\pi(\omega/\nu)$. Thus, for ε small enough and $\lambda \neq k\pi$ $(k = 1, 2, \ldots)$, the mapping T_ε is elliptic too. In other words, stability can change only for:

$$\nu \approx \frac{2\omega}{1}, \frac{2\omega}{2}, \frac{2\omega}{3}, \ldots . \tag{20.3}$$

A sharper computation proves that, for these values of ν (called values of "parametric resonance"; see Appendix 29), the mapping T_ε is hyperbolic at $p = q = 0$ (see Figure 20.4). In other words, the equilibrium of the swing alters (and the swing starts oscillating) if one deflects during an integral number of half-periods of the proper oscillations. This result is well-known empirically.[3]

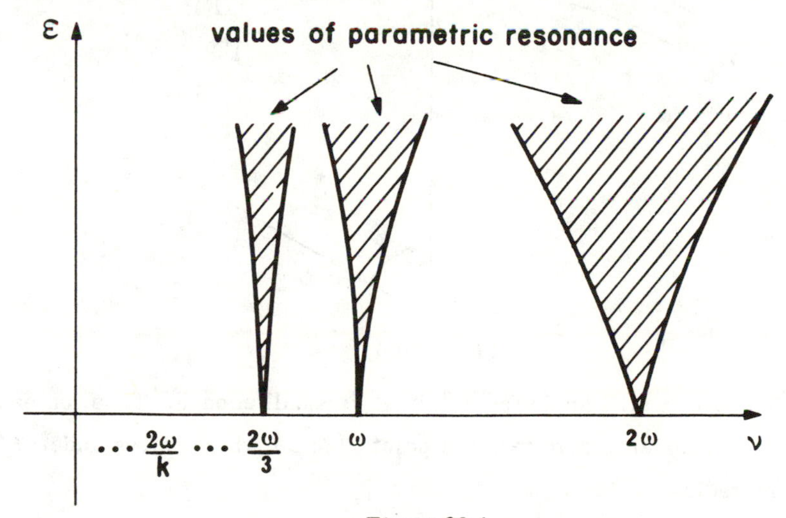

Figure 20.4

[3] Besides, observe that the amplitude of the oscillations of system (19.4) remains small if ε is small enough. For, according to Theorem (19.10), the orbit remains between two curves Γ_ε. This follows from the nonlinearity of our system: the frequency $\lambda(I)$ depends on the amplitude; condition (20.3) is no longer fulfilled since the oscillations do not have time enough to amplify.

THE FIXED POINTS OF THE ITERATES OF T_ε : EXISTENCE 20.5

Now, consider the fixed points of T_ε^n. Let Γ be the invariant curve of T formed with the fixed points of T^n (see §19). We set:

$$\lambda(I) = 2\pi \frac{m}{n}, \qquad \frac{d\lambda}{dI} \neq 0 .$$

On an n-fold iteration of T each point of Γ returns to its original position. This property of T is not retained for a small perturbation $(T \to T_\varepsilon)$. But Poincaré proved, *for* ε *small enough, that* T_ε^n *has* $2kn$ *fixed points close to the curve* Γ. In fact, let us consider two curves that are invariant under T and close to Γ: namely, the curves Γ^+ and Γ^-, with angles of rotation $\lambda^+ > \lambda > \lambda^-$ (Figure 20.6). Thus, the mapping T^n rotates Γ^+ positively and Γ^- negatively.

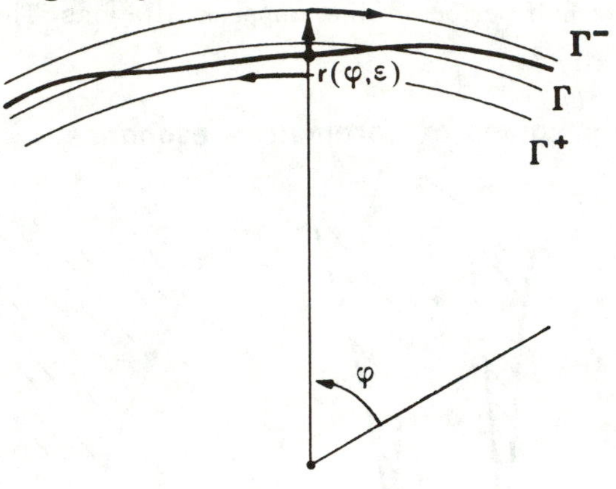

Figure 20.6

This property still holds for T_ε^n if ε is small enough. Thus, on each radius ϕ = constant, there exists a point $r(\phi, \varepsilon)$ which moves under T_ε^n along the radius

$$\phi(T_\varepsilon^n r(\phi, \varepsilon)) = \phi .$$

Moreover, if ε is small enough, the points $r(\phi, \varepsilon)$ $(0 \leq \phi \leq 2\pi)$ form a closed analytical curve R_ε close to Γ. Now, remember that the mapping T_ε^n is canonical and so is area-preserving. Consequently, the image

$T_\varepsilon{}^n R_\varepsilon$ cannot be surrounded by R_ε and, conversely, R_ε cannot be surrounded by $T_\varepsilon{}^n R_\varepsilon$. Thus, R_ε and $T_\varepsilon{}^n R_\varepsilon$ intersect (Figure 20.7). The points of intersection are fixed points of $T_\varepsilon{}^n$, for $T_\varepsilon{}^n$ moves each point of R_ε along its radius $\phi = Ct$. This proves the existence of fixed points of $T_\varepsilon{}^n$ in the neighborhood of Γ.

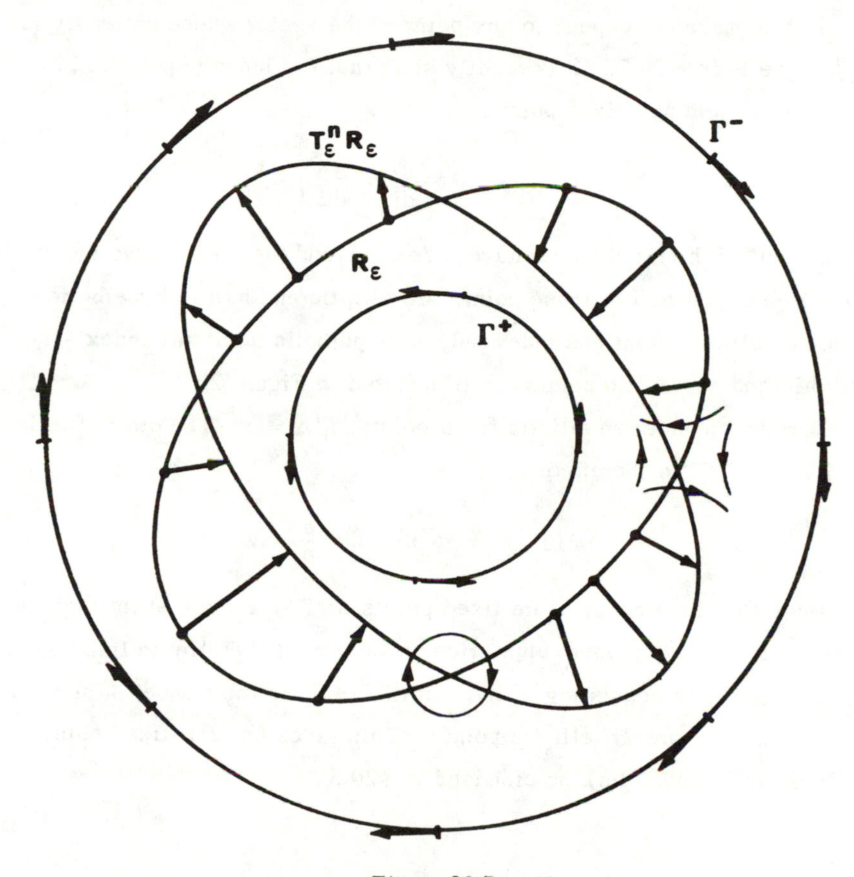

Figure 20.7

FIXED POINTS OF THE ITERATES OF T_ε: CLASSIFICATION 20.8

What is the type of these fixed points? Are they elliptic or hyperbolic? If $\varepsilon = 0$, then they are all parabolic with proper value $\lambda_{12} = 1$. Consequently, for ε small enough, $\lambda_{12} \approx 1$ and the hyperbolic case with reflection is impossible. On the other hand, consider the "radial displacement":

$$\Delta(\phi) = I(T_\varepsilon^n r(\phi)) - I(r(\phi)) .$$

The function $\Delta(\phi)$ vanishes at the fixed points of T_ε^n. In the "generic case" these zeros have multiplicity one ($\Delta' = d\Delta/d\phi \neq 0$). Hence, the zeros at which $\Delta' > 0$ separate the others, and the number of fixed points, is even.

Let us make correspond to any point x the vector whose extremity is $T_\varepsilon^n x$ (see Figure 20.7). It is readily seen that the index (Appendix 27) of this vector field at a fixed point is:

$$\text{Ind} = \text{sign of } \left(\frac{d\lambda}{dI} \cdot \frac{d\Delta}{d\phi}\right) .$$

Thus, half of the fixed points have index $+1$ and the others have index -1. This means that half of these points are elliptic and half of hyperbolic type (an elliptic point has index $+1$, an hyperbolic point has index -1). Elliptic and hyperbolic points are illustrated in Figure 20.7.

Now, consider an elliptic fixed point: $T_\varepsilon^n x = x$. The orbit of x is $x, T_\varepsilon x, \ldots, T_\varepsilon^{n-1} x$, therefore:

$$\phi(T_\varepsilon^l x) \approx \phi(x) + 2\pi l \frac{m}{n} .$$

All points of the orbit of x are fixed points of T_ε^n and are elliptic since they have the same proper value. Hence, the set of the elliptic fixed points splits into orbits consisting of n points. Let k be the number of such orbits, then there are kn elliptic points. This gives us $2kn$ fixed points (elliptic and hyperbolic), as promised in §20.5.

Zones of Instability 20.9

We now turn to the neighborhoods of the above elliptic and hyperbolic fixed points. According to V. Arnold [7] (see also Appendix 28), each "generic" elliptic point is surrounded by closed curves that are invariant under T_ε^n. These curves form "islands" (see Figure 20.10). Each island repeats in miniature the whole structure, with its curves Γ_ε', islands between these curves, and so on. Between these islands and curves Γ_ε, remain zones around the hyperbolic points. In fact,[4] the separatrices of hyperbolic fixed points of the T_ε^n's intersecting each other create an intricate network, as depicted in Figure 20.10. On discovering this, Poincaré wrote ([2], V. 3, Chap. 33, p. 389):

> "One is struck by the complexity of this figure that I am not even attempting to draw. Nothing can give us a better idea of the complexity of the three-body problem and of all problems of dynamics where there is no holomorphic integral and Bohlin's series diverge."

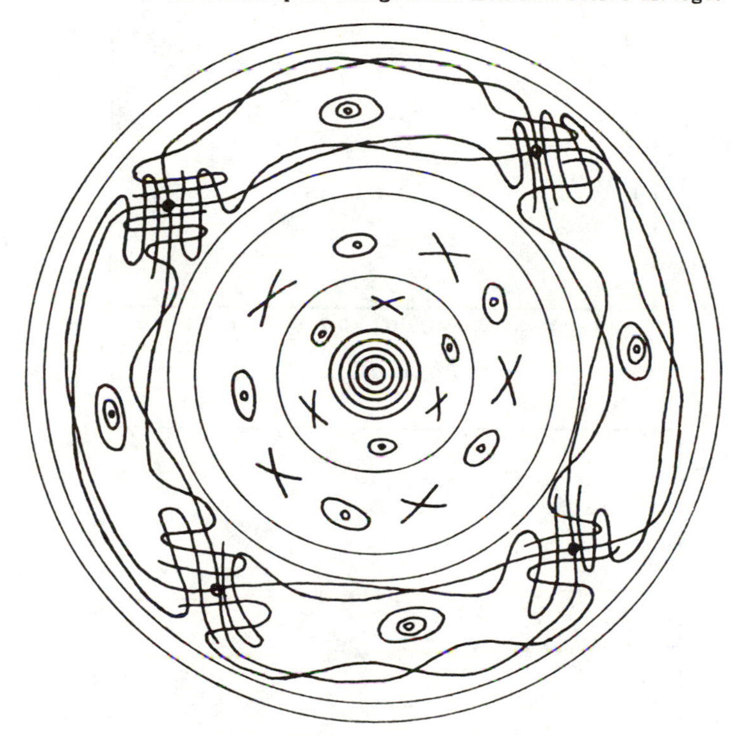

Figure 20.10

The ergodic properties of the motion in zones of instability are unknown. There probably exist systems with singular spectrum and K-systems among the ergodic components.

---

[4] See Poincaré [2], V. Melnikov [1].

Remark 20.11

The above argument does not prove there exist infinitely many elliptic islands for a given $\varepsilon \ll 1$. Poincaré's last geometrical theorem[5] proves that there exist infinitely many fixed points of T_ε^n $(n \to +\infty)$ with index

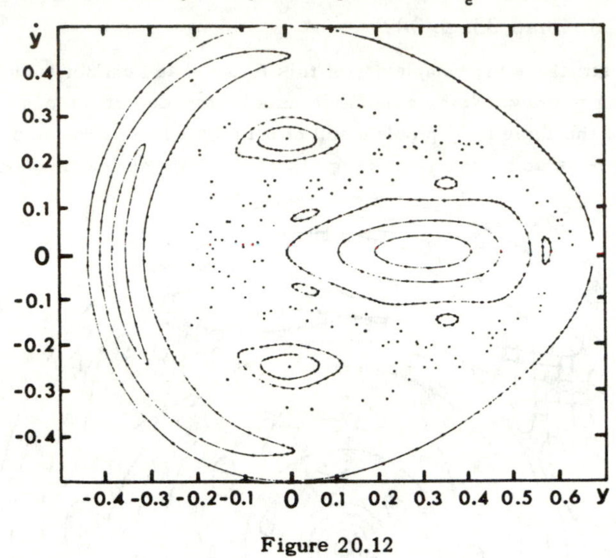

Figure 20.12

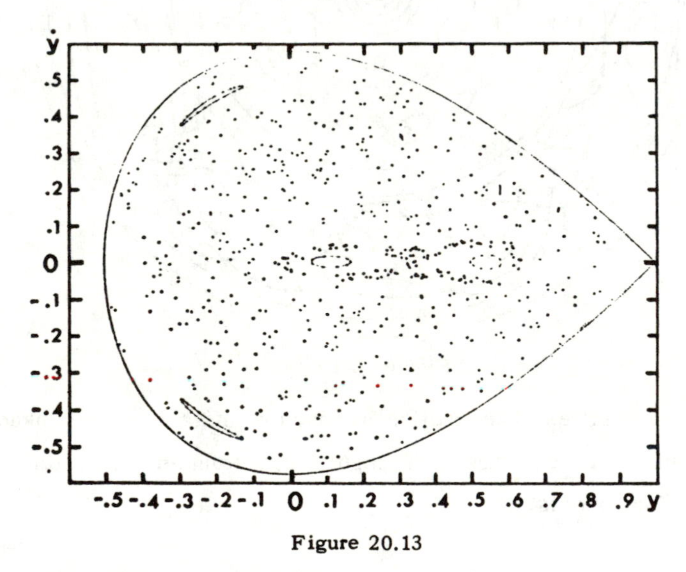

Figure 20.13

---

[5] See Poincaré [3], G. D. Birkhoff [1].

$+1$ inside the annulus located between the invariant curves Γ_ε (Theorem 19.10). Perhaps some of these points are not elliptic but hyperbolic with reflection. Numerical computations[6] seem to support this conclusion.

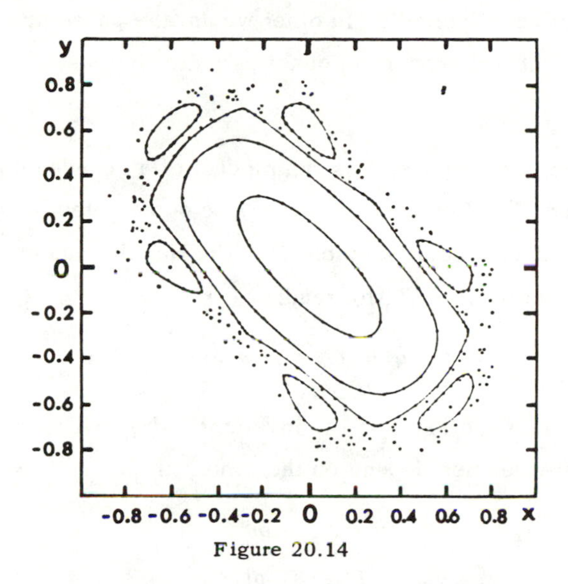

Figure 20.14

We have borrowed Figures 20.12-14 from the work of M. Henon and C. Heiles [1]. They depict the orbits of a mapping of type T_ε computed with an electronic computer. All the points, exterior to the curves, belong to one and the same orbit!

§21. Invariant Tori and Quasi-Periodic Motions

The example we considered in §19 and §20 is a particular case of a situation which occurs for each system close enough to an "integrable" system.

---

[6] See Gelfand, Graev, Sueva, Michailova, Morosov [3]; Ochozimski, Sarychev,... [1]; M. Henon, C. Heiles [1].

(A) INTEGRABLE SYSYEMS 21.1

If one takes a look at the "integrable" problems of Classical Mechanics,[7] one finds that, for all of these problems, bounded orbits are either periodic or quasi-periodic. In other words, the phase-space is stratified into invariant tori supporting quasi-periodic motions.

EXAMPLE 21.2

Assume the phase-space Ω is the product of a bounded domain B^n of $\mathbf{R}^n$ by the torus T^n. Let $p = (p_1, \ldots, p_n)$ be coordinates on B_n and $q = (q_1, \ldots, q_n)$ (mod 2π) coordinates on T^n. The Hamiltonian equations, with Hamiltonian function $H = H_0(p)$, read:

$$\dot{p} = 0, \quad \dot{q} = \omega_0(p), \quad \text{where } \omega_0(p) = \frac{\partial H_0}{\partial p} . \tag{21.3}$$

The motion is quasi-periodic on the invariant tori $p = Ct$, with frequencies $\omega(p)$. Frequencies depend on the torus: if

$$\frac{\partial^2 H_0}{\partial I^2} = \frac{\partial \omega_0}{\partial I} \neq 0,$$

then, on each neighborhood of the torus $p = Ct$, there are invariant tori on which frequencies are independent and orbits everywhere dense (see Appendix 1). There exist other tori on which frequencies are commensurate; they are exceptional, that is they form a set of measure zero. Coordinates (p, q) of $B^n \times T^n$ are called "action-angle" coordinates.

For all integrable systems, it can be shown (Appendix 26) that a certain $(2n-1)$-dimensional hypersurface divides the phase-space into invariant domains each of which is stratified into invariant n-dimensional manifolds. If the domain is bounded, these manifolds are tori supporting quasi-periodic motions. The action-angle coordinates can be introduced into such a domain, thus, the system can be described by (21.3).

---

[7] For instance, the motion of a free point along a geodesic on the surface of a triaxial ellipsoid or a torus (see (1.7) and Appendix 2), a heavy solid body (Euler, Lagrange, and Kovalewskaia cases).

(B) SYSTEMS CLOSE TO INTEGRABLE SYSTEMS 21.4

Now, we assume that the Hamiltonian function is perturbed:

$$H = H_0(p) + H_1(p, q), \quad H_1(p, q + 2\pi) = H_1(p, q) \ll 1 ,$$

the "perturbation" H_1 being "small enough." The Hamiltonian equations are then:

$$\text{(21.5)} \qquad \dot{p} = -\frac{\partial H_1}{\partial q}, \quad \dot{q} = \omega_0(p) + \frac{\partial H_1}{\partial p} .$$

For most initial data, A. N. Kolmogorov [6] proved that the motion remains quasi-periodic (see Theorem 21.7). Consequently, (21.5) is not ergodic on the "energy surface" $H = Ct$ and, among the ergodic components, there are components with discrete spectrum, the complement of which has small Lebesgue measure as H_1 is small.

Assume that the function $H(p)$ is analytic in a complex domain $[\Omega]$ of the phase-space:

$$\Omega(\Re p, \Re q, \in \Omega, \ |\Im p| < p, \ |\Im q| < p) .$$

Assume also that the unperturbed system is nondegenerate:

$$\text{(21.6)} \qquad \mathrm{Det}\left|\frac{\partial \omega_0}{\partial p}\right| = \mathrm{Det}\left|\frac{\partial^2 H_0}{\partial p^2}\right| \neq 0 .$$

Select an incommensurate[8] frequency-vector $\omega = \omega^*$. The equations of the invariant torus $T_0(\omega^*)$ of the unperturbed system (21.3) are $p = p^*$, where $\omega_0(p^*) = \omega^*$. Thus, the system (21.3) has frequencies ω^* on $T_0(\omega^*)$.

THEOREM 21.7

If H_1 is small enough, then for almost[9] all ω^*, there exists an invariant torus $T(\omega^*)$ of the perturbed system (21.5) and $T(\omega^*)$ is close to $T_0(\omega^*)$. To be precise:

---

[8] That is, $(\omega, \mathbf{k}) \neq 0$ for all integers $\mathbf{k}$.

[9] All, except for a set of Lebesgue measure zero.

For all $K > 0$ *there exist* $\varepsilon > 0$ *and a mapping* $p = p(Q)$, $q = q(Q)$ *of an abstract torus* $T = \{Q \pmod{2\pi}\}$ *into* Ω *such that, according to* (21.5)

$$\dot{Q} = \omega^* ,$$

and:

$$|p(Q) - p^*| < K, \quad |q(Q) - Q| < K,$$

if:

$$|H_1| < \varepsilon = \varepsilon(K, \omega^*, H_0, \Omega, \rho) > 0$$

holds in $[\Omega]$.

Moreover, the tori $T(\omega^*)$ form a set of positive measure whose complement set has a measure that tends to zero as $|H_1| \to 0$. Proof of Theorem (21.7) is found in Arnold [5].

The behavior of the orbits emanating from this complement is not well known. If our system has two degrees of freedom ($n = 2$) then the phase-space Ω has dimension four, and the two-dimensional invariant tori that are found divide the three-dimensional manifold H = constant. The domains of the complement set are toric *annuli* between these invariant tori (see Figure 19.11). For $n > 2$, the n-dimensional invariant tori do not divide the $(2n-1)$-dimensional manifold H = constant and those orbits that do not belong to the tori $T(\omega^*)$ can travel very far along $H = h$ (see §23).

(C) APPLICATIONS AND GENERALIZATIONS 21.8

Theorem (21.7) applies to the *motion of a free point along a geodesic on a convex surface close to an ellipsoid or a surface of revolution*. This theorem allows one to prove the *stability in the plane restricted circular three-body problem.*[10] One can also deduce the *stability of the fast rotations of a heavy asymmetric solid body.*[11]

But this theorem does not apply if the unperturbed motion has fewer frequencies than the perturbed motion (degenerate case) for, in this case,

---

[10] A. N. Kolmogorov [7].
[11] V. I. Arnold [5].

condition (21.6) does not hold:

$$\operatorname{Det} \frac{\partial^2 H_0}{\partial p^2} \equiv 0 .$$

The cases of "limiting degeneracy" of the oscillation theory (points of equilibrium, periodic motions) also require a particular study. In that direction we mention some results that generalize Theorem (21.7).

V. I. Arnold [7] proved the *stability of positions of equilibrium and of periodic motions* of systems with two degrees of freedom in the general elliptic case. As a corollary, A. M. Leontovich [1] deduced the *stability of the Lagrange periodic solutions for the reduced problem of the three-body (plane and circular).*

V. Arnold [8], [9], [10], studied the generation of new frequencies from the perturbation of degenerate systems. As a corollary, one obtains the *perpetual adiabatic invariance of the action* for a slow periodic variation of the parameters of a nonlinear oscillatory system with one degree of freedom, and also that a *"magnetic trap" with an axial-symmetric magnetic field can perpetually retain charged particles.*

Finally, *quasi-periodic motions in the n-body problem have been found.* If the masses of n planets are small enough in comparison with the mass of the central body, the motion is quasi-periodic for the majority of initial conditions for which the eccentricities and inclinations of the Kepler ellipses are small. Further, the major semiaxes perpetually remain close to their original values, and the eccentricities and inclinations remain small (see V. Arnold [4]).

On the other hand, J. Moser [1], [5] generalized Theorem (21.7). Moser abandons the requirement of analyticity of the Hamiltonian and *substitutes instead the requirement that several hundred derivatives exist.* For instance, for systems with two degrees of freedom, it is sufficient that 333 derivatives exist!

(D) INVARIANT TORI OF CANONICAL MAPPINGS 21.9

Theorem (21.7) can be reformulated by using the construction of the "surfaces of section" of Poincaré-Birkhoff. Assume that, in Equations (21.3), the first component ω_1 of ω is nonvanishing. Consider a submanifold Σ^{2n-2} of the phase-space Ω^{2n} whose equations are: $q_1 = 0$, $H = h =$ constant. The orbit $x(t)$ of (21.5) through a point x on Σ^{2n-2} will, as t increases from zero, return to Σ^{2n-2} and will cut Σ^{2n-2} in a uniquely determined point Ax (Figure 19.11). If the perturbation H_1 is small enough and $\omega_1(p) \neq 0$, the mapping $A\colon \Sigma^{2n-2} \to \Sigma^{2n-2}$ is well defined in a neighborhood of the $(n-1)$-dimensional torus: $p = Ct$, $q_1 = 0$. Since

$$\frac{\partial H}{\partial I_1} = \omega_1 \neq 0 ,$$

then $p_2, \ldots, p_n;\ q_2, \ldots, q_n \pmod{2\pi}$ are "action-angle" coordinates in this neighborhood. The mapping A is canonical (see Appendix 31).

Now, consider the unperturbed system ($H_1 = 0$). According to (21.3), the map A may be written as follows:

$$\text{(21.10)} \qquad A\colon\ p, q \to p, q + \omega(p);\ \omega_k(p) = 2\pi \frac{\omega_k}{\omega_1} \qquad (k = 2, \ldots, n).$$

In other words, each torus $p = Ct$ is invariant and rotates through $\omega(p)$ under the mapping A.

If the perturbation H_1 is small, then the corresponding canonical mapping $\mathcal{A}\colon \Sigma^{2n-2} \to \Sigma^{2n-2}$ is close to (21.10). The $(n-1)$-dimensional invariant tori of $\mathcal{A}$ are, obviously, similar to the n-dimensional invariant tori of (21.5) and there is a theorem, similar to Theorem (21.7), for mappings (see Theorem 21.11).

Let Ω again be the phase-space p, q:

$$\Omega = T^n \times B^n, \quad B^n = \{p\}, \quad T^n = \{q \pmod{2\pi}\} .$$

Assume that $\mathrm{B}\colon\ p, q \to p'(p, q), q'(p, q)$ is a global canonical mapping, that is:

$$\oint_\gamma p\,dq = \oint_{\mathrm{B}\gamma} p\,dq ,$$

for any closed curve γ of Ω (see Appendix 33). Then, assume that the functions $p'(p, q)$, $q'(p, q) - q$ are analytic in the complex neighborhood $[\Omega]$ of Ω:

$$\mathfrak{R}(p), \mathfrak{R}(q) \in \Omega; \quad |\mathfrak{J}p|, |\mathfrak{J}q| < \rho .$$

Let $\mathbf{A}$: $p, q \to p, q + \omega(p)$ be the canonical mapping defined by an analytic function $\omega(p)$ in $[\Omega]$, and $T_0(\omega^*)$ the torus $p = p^*$, $\omega(p^*) = \omega^*$ that is invariant under $\mathbf{A}$.

THEOREM 21.11

Suppose, that the system is nondegenerate, that is the jacobian of ω does not vanish identically.

If $\mathbf{B}$ *is close enough to the identity, then, for almost* [12] *all* ω^*, *there exists a torus* $T(\omega^*)$ *that is invariant under* $\mathbf{BA}$ *and close to* $T_0(\omega^*)$.

To be precise, to any $K > 0$ corresponds an $\varepsilon > 0$ and a mapping $\mathbf{D}$: $T \to \Omega$, $p = p(Q)$, $q = q(Q)$, of an abstract torus $T = \{Q \pmod{2\pi}\}$ into Ω, such that:

$$\mathbf{D}(Q + \omega^*) = \mathbf{B} \cdot \mathbf{A} \cdot \mathbf{D}(Q) ,$$

$$\begin{array}{ccccc} & \xrightarrow{\mathbf{A}} & & \xrightarrow{\mathbf{B}} & \\ \mathbf{D} \uparrow & & & & \uparrow \\ Q & & \longrightarrow & & Q + \omega^* \end{array} ,$$

and:

$$|p(Q) - p^*| < K, \qquad |q(Q) - Q| < K ,$$

provided that:

$$|p' - p| + |q' - q| < \varepsilon = \varepsilon(K, \omega^*, \mathbf{A}, \Omega, \rho) > 0$$

holds in $[\Omega]$. Moreover the tori $T(\omega^*)$ form a set of positive measure whose complement set has measure tending to zero as $|p' - p| + |q' - q| \to 0$.

Theorem (19.10) is a direct corollary of Theorem (21.11): $n = 1$. Theorem (21.11) has been known since 1954, though its proof was never published. J. Moser [1] gave a proof for mappings of the plane ($n = 1$). This proof makes use of the topology $\mathbf{R}^2$. Appendix 34 gives a proof for arbitrary n: the topological part is reduced to the technique of generating functions of global canonical transformations (see Appendix 33).

---

[12] All, except for a set of Lebesgue measure zero.

(E) COMPARISON OF THEOREM (21.7) WITH THEOREM (21.11) 21.12

Whether each analytic canonical transformation close to A can be constructed by using a section of a suitable Hamiltonian system is unknown. Thus, Theorem (21.11) cannot be deduced from Theorem (21.7).

Even if one restricts to the Hamiltonian system (21.5), Theorems (21.7) and (21.11) are nonequivalent. In fact, according to (21.6) and (21.10), the nondegeneracy conditions of Theorems (21.7) and (21.11):

$$\text{Det} \left| \frac{\partial \omega}{\partial p} \right| \neq 0 , \qquad \text{Det} \left| \frac{\partial \omega}{\partial p} \right| \neq 0$$

may be written in terms of the unperturbed Hamiltonian function H_0 as follows:

$$\text{Det} \left| \frac{\partial^2 H_0}{\partial p^2} \right| \neq 0, \qquad \text{Det} \begin{vmatrix} \frac{\partial^2 H_0}{\partial p^2} & \frac{\partial H_0}{\partial p} \\ \frac{\partial H_0}{\partial p} & 0 \end{vmatrix} \neq 0 . \tag{21.13}$$

These conditions (21.13) are clearly independent. Each of them is sufficient to ensure that there exist invariant tori. The second one ensures, additionally, that there exists such tori on *each* "energy surface"; this implies the stability (see Figure 19.11) for systems with two degrees of freedom ($n = 2$ for Theorem 21.7, $n = 1$ for Theorem 21.11).

In most applications, both of the conditions (21.13) are simultaneously fulfilled or invalid.

§ 22. Perturbation Theory

We turn next to the asymptotic theory, that is we restrict our study to the behavior of orbits for $0 < t < 1/\varepsilon$, ε being the magnitude of the perturbation. In contrast, non-Hamiltonian systems can be considered.

(A) AVERAGING METHOD[13] 22.1

Let $T^k = \{\phi = (\phi_1, \dots, \phi_k) \pmod{2\pi}\}$ be the k-dimensional torus and $B^l = \{I = (I_1, \dots, I_l)\}$ a bounded domain of $\mathbf{R}^1$. In the phase-space $\Omega = T^k \times B^l$, consider the unperturbed system:

$$\dot{\phi} = \omega(I), \quad \dot{I} = 0, \quad \omega = (\omega_1, \dots, \omega_k) . \tag{22.2}$$

It is, obviously, a generalization of system (21.3): each torus I = constant is invariant and, if the frequencies ω are incommensurate on this torus T, then the orbits $\phi(t)$ are everywhere dense on T. In such a case, the motion (22.2) is called quasi-periodic on the torus T. If the frequencies are commensurate, then the closure of an orbit is a k-dimensional torus $(k < n)$ (resonance).

Consider next the perturbed system that generalizes (21.5):

$$\begin{cases} \dot{\phi} = \omega(I) + \varepsilon f(I, \phi) \\ \dot{I} = \varepsilon F(I, \phi) \end{cases}, \text{ where } \begin{cases} f(I, \phi + 2\pi) \equiv f(I, \phi) \\ F(I, \phi + 2\pi) \equiv F(I, \phi), \end{cases} \quad \varepsilon \ll 1 . \tag{22.3}$$

Of course, for $t \approx 1$, the evolution $|I(t) - I(0)| \sim \varepsilon \ll 1$. Notable effects, of order 1, of the evolution appear only after a long enough time: $t \sim 1/\varepsilon$.

Perturbation Theory proceeds to study the perturbed system as follows. Let $\bar{F}(I)$ be the mean:

$$\bar{F}(I) = (2\pi)^{-k} \cdot \oint \cdots \oint F(I, \phi) \, d\phi_1 \cdots d\phi_k \quad .$$

One considers the "averaged system," or "system of evolution":

$$\dot{J} = \varepsilon \cdot \bar{F}(J) . \tag{22.4}$$

For $\varepsilon \ll 1$, one expects that:

$$|I(t) - J(t)| \ll 1 \text{ for } 0 < t < \frac{1}{\varepsilon} \tag{22.5}$$

where $I(t)$, $\phi(t)$ is a solution of (22.3) and $J(t)$ is the solution of (22.4) with initial data: $J(0) = I(0)$.

---

[13] This method goes back to Lagrange, Laplace, and Gauss, who used it in Celestial Mechanics.

Now the problem arises as follows: *what relations exist, for* $0 < t < \frac{1}{\varepsilon}$, *between the perturbed motion* $I(t)$ *and the "motion of evolution"* $J(t)$? *Does the inequality* (22.5) *hold?*

For the simplest periodic motions ($k = 1$) it is readily proved (see Appendix 30 and Bogolubov and Mitropolski [1]) that if $\omega \neq 0$ then:

$$|I(t) - J(t)| < C \cdot \varepsilon, \text{ for } 0 < t < \frac{1}{\varepsilon} \cdot$$

But the situation becomes more complicated as the number k of frequencies increases, even for $k = 2$.

(B) A COUNTER-EXAMPLE 22.6

Assume $k = l = 2$, $a > 1$ and consider the system:

$$\dot{\phi}_1 = I_1, \quad \dot{\phi}_2 = I_2, \quad \dot{I}_1 = \varepsilon, \quad \dot{I}_2 = \varepsilon a \cos(\phi_1 - \phi_2) .$$

Of course, the system of evolution is:

$$\dot{J}_1 = \varepsilon, \quad \dot{J}_2 = 0$$

(corresponding to small arrows on Figure 22.7). Consider the following initial data:

$$I_1 = I_2 = J_1 = J_2 = 1, \quad \phi_1 = 0, \quad \phi_2 = \text{arcos}\, \frac{1}{a} .$$

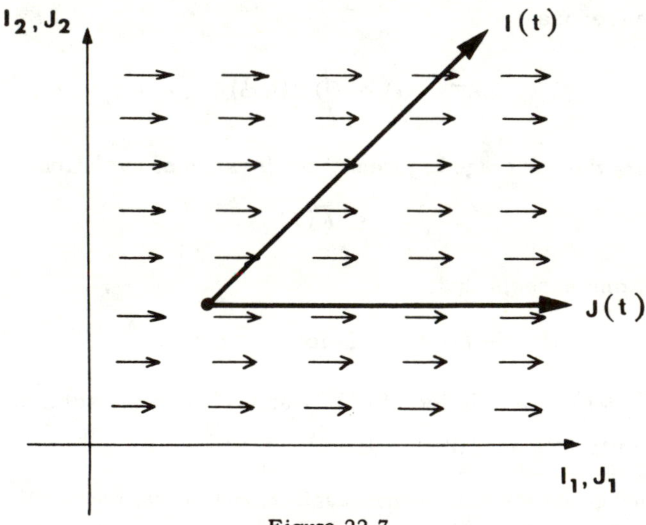

Figure 22.7

Then:

$$I_1(t) = I_2(t) = 1 + \varepsilon t, \ \ J_1(t) = 1 + \varepsilon t, \ \ J_2(t) = 1 .$$

Thus,

$$|I(1/\varepsilon) - J(1/\varepsilon)| = 1 .$$

In other words, after the interval of time $1/\varepsilon$, the averaged motion loses any relation with the real motion which remains locked in by the resonance $\omega_1 = \omega_2$.

(C) MATHEMATICAL FOUNDATIONS OF THE AVERAGING METHOD 22.8

There exist, at least, four distinct approaches to the problem of the mathematical foundations of averaging method. All four lead to rather modest results.

(1) *The neighborhoods of particular solutions* (for example, positions of equilibrium $\bar{F} = 0$) of the averaged system can be fairly well studied. For instance, there exist attracting tori of (22.3) which correspond to the attracting points of system (22.4). Stability (for $0 < t < \infty$) obviously holds in the neighborhood of such a torus. N. N. Bogolubov [2], J. Moser [2], [5], and I. Kupka [1] proved that attracting tori still exist for perturbed systems.

This approach does not apply to Hamiltonian systems because attracting points do not exist according to the Liouville theorem (see 1.10).

(2) One can study the relations between $I(t)$ and $J(t)$ *for most (in the sense of measure theory) initial data,* neglecting points that correspond to resonance. For instance, Anosov [3] and Kasuga [1] proved theorems of the following type:

Let $R(\varepsilon, \rho)$ be the subset of Ω of the initial data such that

$$|I(t) - J(t)| > \rho$$

for certain $0 < t < 1/\varepsilon$. Then, $\lim_{\varepsilon \to 0}$ measure $R(\varepsilon, \rho) = 0$ for all $\rho > 0$.

This approach allows one to obtain similar results for systems much more general than (22.3); whence its weakness: estimates of the measure of $R(\varepsilon, \rho)$ are not realistic and one has no information concerning the motion in $R(\varepsilon, \rho)$.

(3) One can study *passages through states of resonance.*

(4) *One restricts oneself to Hamiltonian systems* to obtain more information.

(D) Passage through States of Resonance 22.9

Let us begin with an example:

$$\dot{\phi}_1 = I_1 + I_2, \quad \dot{\phi}_2 = I_2, \quad \dot{I}_1 = \varepsilon, \quad \dot{I}_2 = \varepsilon \cos(\phi_1 - \phi_2).$$

The averaged equations are (see Figure 22.10):

$$\dot{J}_1 = \varepsilon, \quad \dot{J}_2 = 0\,.$$

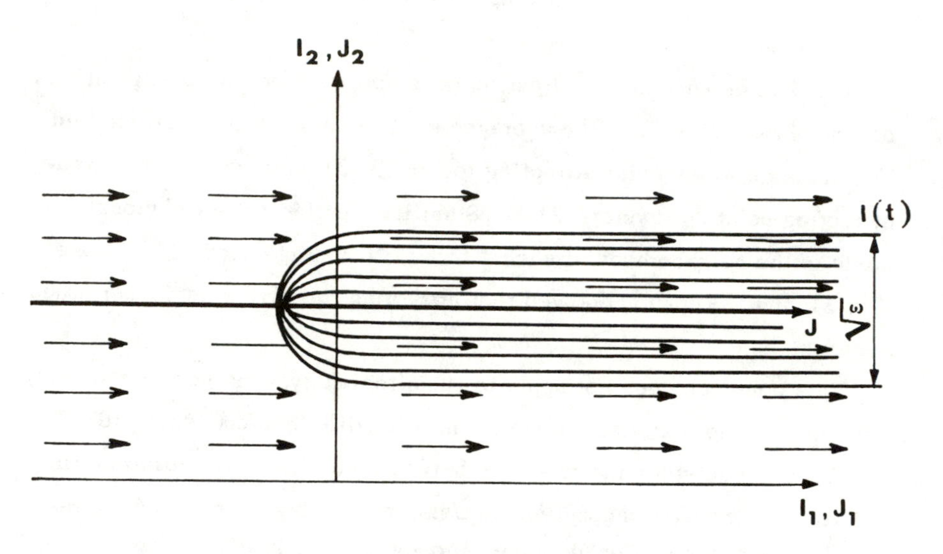

Figure 22.10

Consider the initial data that correspond to resonance $\omega_1 = \omega_2$:

$$\phi_1(0) = \phi_2(0) = I_1(0) = I_2(0) - 1 = 0\,.$$

The system is easily integrated:

$$|I(t) - J(t)| = |I_2(t) - 1| = \sqrt{2\varepsilon} \int_0^{\tau} \cos x^2 \cdot dx\,, \quad \tau = \sqrt{\varepsilon/2t}\,.$$

For $t = 1/\varepsilon$, obviously,

$$\left| I\left(\frac{1}{\varepsilon}\right) - J\left(\frac{1}{\varepsilon}\right) \right| = C \cdot \sqrt{\varepsilon} .$$

Thus, the passage through the resonance $\omega_1 = \omega_2$ disperses the bundle of orbits $I(t)$, $\phi(t)$, which in the beginning differ only by phases $\phi(0)$. The scattering of I_2 after going through the resonance is of the order of $\sqrt{\varepsilon}$ (see Figure 22.10). For a general system (22.3) with two frequencies ($k = 2$), one obtains[14] the following theorem:

If the quantity:

$$A(I, \phi) = \left(\frac{\partial \omega_1}{\partial I} F\right) \omega_2 - \left(\frac{\partial \omega_2}{\partial I} F\right) \omega_1$$

does not vanish in Ω, then we have the estimate:

$$|I(t) - J(t)| < C \cdot \sqrt{\varepsilon} \cdot \ln^2 \left(\frac{1}{\varepsilon}\right) \quad \textit{for all } 0 < t < \frac{1}{\varepsilon} . \tag{22.11}$$

Condition $A \neq 0$ means that the system cannot remain locked in at any resonance: (22.3) implies

$$\frac{d(\omega_1/\omega_2)}{dt} \neq 0 .$$

In example (22.6) condition $A \neq 0$ is violated:

$$A(I, \phi) = I_2 - I_1 a \cos(\phi_1 - \phi_2)$$

changes sign at $I_1 = I_2$ if $a > 1$. This example shows that condition $A \neq 0$ cannot be replaced by an analogous condition for the averaged system.

The idea used in proving (22.11) is that the scattering produced by each resonance is of the order $C\sqrt{\varepsilon}$ and that, among the infinitely many resonances $\omega_1/\omega_2 = m/n$, only the greatest

$$\ln^2 \frac{1}{\varepsilon} \qquad (m, n < \ln \frac{1}{\varepsilon})$$

produces notable effects.

---

[14] V. I. Arnold [12].

Passage through resonances for systems with more than two frequencies ($k > 2$) has not been studied.

(E) Evolution of Hamiltonian Systems 22.12

Next, apply the averaging method to Hamiltonian systems (21.5). If condition (21.6) of nondegeneracy holds, then most of the unperturbed orbits are ergodic on tori p = constant. Thus, it is satisfactory to write this system in the form (22.3), with $I = p$, $\phi = q$, $k = l = n$:

$$\begin{cases} \dot{\phi} = \omega(I) + \varepsilon \dfrac{\partial H_1}{\partial I} \\ \dot{I} = -\varepsilon \dfrac{\partial H_1}{\partial \phi}, \end{cases}$$

where

$$H = H_0 + \varepsilon H_1, \qquad \omega = \frac{\partial H_0}{\partial p} .$$

The averaged system is $\dot{J} = 0$, for

$$\bar{F}(J) = -(2\pi)^{-k} \cdot \oint \cdots \oint \frac{\partial H_1(I, \phi)}{\partial \phi} \, d\phi_1 \cdots d\phi_n \equiv 0 .$$

In other words, *there is no evolution for nondegenerate Hamiltonian systems*: J = constant.

Theorem (21.7) of conservation of quasi-periodic motions rigorously establishes this conclusion. In fact, Theorem (21.7) implies that:

$$|I(t) - J(t)| < K \text{ for all } t \in \mathbb{R} \text{ and } \varepsilon < \varepsilon_0(K)$$

(for all initial data if $n = 2$, and

$$\operatorname{Det} \begin{pmatrix} \dfrac{\partial^2 H_0}{\partial I^2} & \dfrac{\partial H_0}{\partial I} \\ \dfrac{\partial H_0}{\partial I} & 0 \end{pmatrix} \neq 0,$$

and for most initial data in the general case). This illuminates the part that conditions of conservation play in Theorem (21.7): they prevent evolution.[15] In the same way, evolution is prohibited in Theorem (21.11), for the mapping is *globally* canonical.

On the other hand, one also understands the part that the condition of nondegeneracy plays. In fact, in case of degeneracy, the generic orbit of the unperturbed system is ergodic on k-dimensional tori (k *strictly* inferior to n). In such a case, the algorithm of perturbation theory allows one to predict the averaging on the k-dimensional torus. Hence evolution becomes possible, even for canonical systems.

---

[15] This particular feature of canonical systems already comes out from simple examples. Let us consider the following perturbations of a center (Figure 22.13):

$$\begin{cases} \dot{x} = y + \varepsilon_1 \\ \dot{y} = -x + \varepsilon_2 \end{cases} \qquad \begin{cases} \dot{x} = y \\ \dot{y} = -x - \varepsilon y \end{cases} .$$

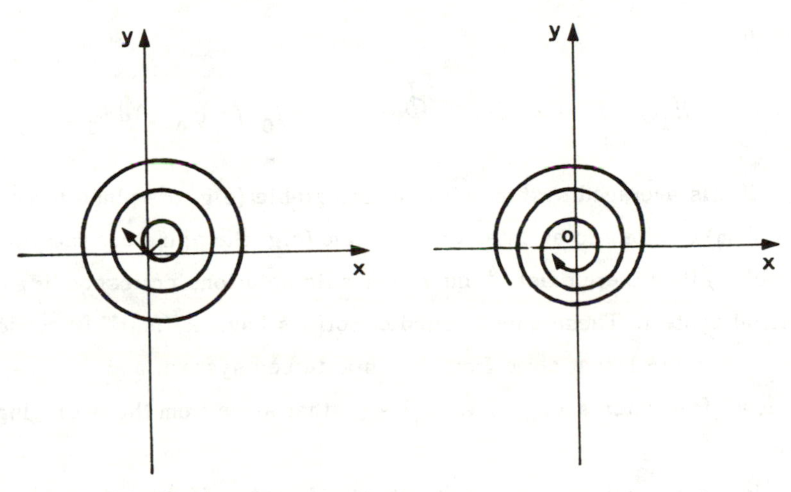

Figure 22.13

The first perturbation, which is canonical, moves the orbits along a direction which is orthogonal to the perturbation, and does not give rise to evolution. The second perturbation, which is not canonical, gives rise to an evolution to zero. Volume-preserving examples with evolution can be constructed in the four-dimensional space x, y, z, u:

$$\dot{x} = y, \quad \dot{y} = -x - \varepsilon y, \quad \dot{z} = u, \quad \dot{u} = -z + \varepsilon u .$$

EXAMPLE 22.14

Consider the Hamiltonian system:

$$H = H_0 + \mathrm{M}H_1(I_1, \ldots, I_n, \phi_1, \ldots, \phi_n), \quad I \in B^n,$$

$$\phi \in T^n = \{(\phi_1, \ldots, \phi_n) \bmod 2\pi\}$$

$$H_0 = H_0(I_1, \ldots, I_k), \qquad \frac{\partial^2 H_0}{\partial I^2} \neq 0 .$$

This system has the form (22.3), with $k < n$, $l = 2n-k$, and the averaged system is:

$$\begin{cases} \dot{J}_0 = 0 & J_0 = (J_1, \ldots, J_k), \ \phi_0 = (\phi_1, \ldots, \phi_k) (\bmod 2\pi) \\ \dot{J} = -\varepsilon \cdot \dfrac{\partial \bar{H}_1}{\partial \psi} & J = (J_{k+1}, \ldots, J_n) \\ \dot{\psi} = \varepsilon \dfrac{\partial \bar{H}_1}{\partial J} & \psi = (\psi_{k+1}, \ldots, \psi_n) \ (\bmod 2\pi) , \end{cases}$$

where

$$\bar{H}_1(J_0, J, \psi) = (2\pi)^{-k} \oint \cdots \oint H_1(J_0, J;\ \phi_0, \psi)\, d\phi_0 .$$

If this averaged system is either integrable (e.g. the plane three-body problem) or close to an integrable system (e.g. the planetary many-body problem) then there exist [16] quasi-periodic solutions corresponding to the initial system. These quasi-periodic motions have k "fast" frequencies $(\omega_1, \ldots, \omega_k) \sim 1$ that come from the unperturbed system, and $l = n-k$ "slow" frequencies $(\omega_{k+1}, \ldots, \omega_n) \sim \varepsilon$ that arise from the averaging system.

In the general case, when the averaged system is not integrable, the relation between the solutions of the perturbed and the averaged systems is still unknown even for $0 < t < 1/\varepsilon$. The only known results arise from approaches 2 and 3 (22.8). Moreover, observe that, even for nondegenerate systems, we need a study of the motion in the zones of instability

[16] V. Arnold [10], [4].

(complement set of the invariant tori) for $n > 2$, at least for $t \sim 1/\varepsilon$ (or $t \sim 1/\varepsilon^m$). In such a zone, one can probably find,[17] $(n-1)$-dimensional invariant tori of "elliptic" or "hyperbolic" type that generalize, in arbitrary dimension, periodic motions of §20. If $n > 2$, recall that n-dimensional invariant tori do not divide the $(2n-1)$-dimensional energy level H = constant. Consequently, the "separatrices" of the "hyperbolic" tori can travel very far along H = constant, producing instability. The next section is devoted to the study of a similar mechanism of instability.

§23. Topological Instability and Whiskered Tori

We give next an example[18] (see 23.10) of an Hamiltonian system that satisfies conditions of Theorems (21.7) and (21.11), but that is topologically unstable: $|I(t) - J(t)|$ is unbounded for $-\infty < t < \infty$. According to Theorems (21.7) and (21.11), this system is stable for most initial data (the corresponding motions are quasi-periodic). The secular changes of $I(t)$ have the velocity $\exp(-1/\sqrt{\varepsilon})$ and consequently cannot be dealt with by any approximation of the classical theory of perturbations.

We first introduce some definitions.

(A) THE WHISKERED TORI 23.1

Assume that in the phase-space of the dynamical system there is an invariant torus T and on it a quasi-periodic motion with everywhere dense orbits.

We shall call T a *whiskered torus* if T is a connected component of the intersection of two invariant open manifolds: $T = M^+ \cap M^-$, where

---

[17] One can find motivation in V. Arnold [14]. Since this was written, the proof was given independently by V. K. Melnikov [2], J. Moser [5], and G. A. Krasinskii [1].

[18] Example (23.10) is rather artificial, but we believe that the mechanism of "transition chains" which guarantees that instability in our example is also applicable to the generic case (for instance, to the three-body problem).

all the orbits on *arriving whisker* M^- approach T as $t \to +\infty$, and on the *departing whisker* M^+ all the orbits approach T as $t \to -\infty$:

$$\lim_{t \to -\infty} |x(t) - T| = 0 \quad \text{for } x(0) \in M^+$$

$$\lim_{t \to +\infty} |x(t) - T| = 0 \quad \text{for } x(0) \in M^-.$$

For instance, the torus T^k: $x = y = z = 0$ in the system:

$$\dot{x} = \lambda \cdot x, \quad \dot{y} = -\mu \cdot y, \quad \dot{z} = 0, \quad \dot{\phi} = \omega \tag{23.2}$$

($\lambda, \mu > 0$, $\phi (\text{mod } 2\pi) \in T^k$, ω incommensurate) defined in the space $\mathbf{R}^{l+} \times \mathbf{R}^{l-} \times \mathbf{R}^{l_0} \times T^k$ has a $(l_+ + k)$-dimensional whisker M^+ $(y = z = 0)$ and a $(l_- + k)$-dimensional whisker M^- $(x = z = 0)$.

(B) THE TRANSITION TORI 23.3

Let M be a smooth submanifold of space X. We shall say that the subset $\Omega \subset X$ *obstructs* the manifold M at the point $x \in M$ if every manifold N that is transverse to M at x is intersected by Ω. For instance,[19] a spiral Ω which winds onto a limit cycle M obstructs M (see Figure 16.4, Chapter 3). If the whiskered torus T has the property that the images of an arbitrary neighborhood U of an arbitrary point ξ of one of its arriving whiskers M^- obstruct the departing whisker M^+ at an arbitrary point η of M^+, then the torus will be said to be a *transition torus* (see Figure 23.4).

LEMMA 23.5

The torus $x = y = z = 0$ *in system* (23.2) *is a transition torus.*

Proof:

We set $\xi = (0, y_0, 0, \phi_0)$, $\eta = (x_1, 0, 0, \phi_1)$. The ω's being incommensurate, there exists a sequence t_i, $t_i \to +\infty$, such that the distance from $\phi_0 + \omega t_i$ to ϕ_1 tends to zero.

Consider the part V of U whose equation is $y = y_0$. By $\Omega = \bigcup_{t>0} U(t)$

---

[19] Articles by Sitnikov [1] and A. Leontovich [1] are based on this fact.

we denote the set of all points of all the orbits emanating from U. Then Ω contains the set of all the images $g_{t_i}V$, where g_t is the transformations

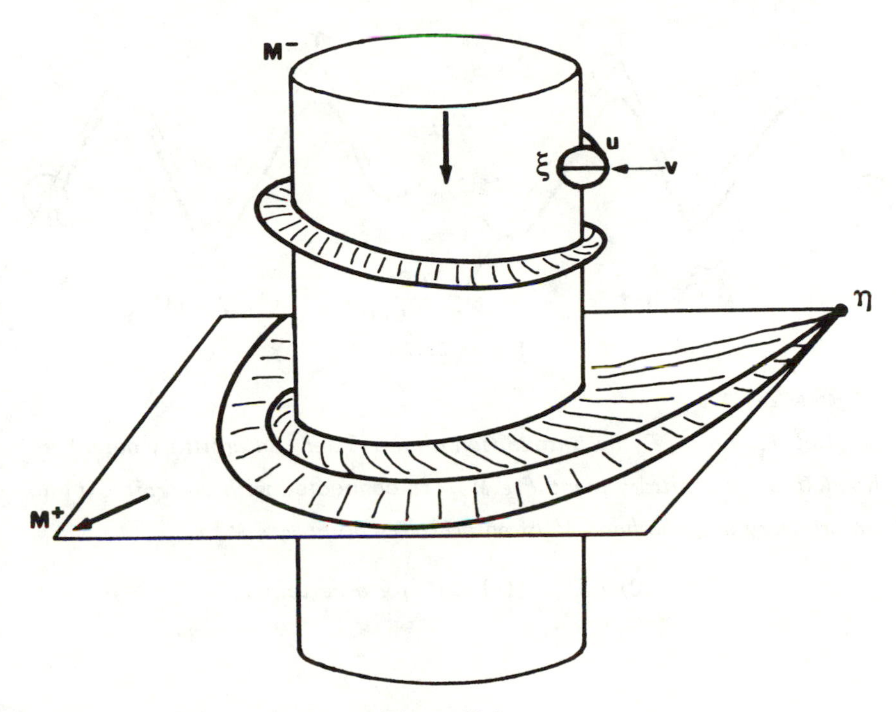

Figure 23.4

group defined by (23.2). For t_i large enough, these images $g_{t_i}V$ intersect the neighborhood of η (because $\lambda > 0$). The intersections have equations:

$$y = y_i, \quad y_i = e^{-\mu t_i} \cdot y_0 \to 0 \, .$$

Thus Ω contains the set of all the surfaces $g_{t_i}V$ that are parallel to M^+ and converge to M^+. These surfaces already obstruct M^+ at η; this proves Lemma (23.5).

(C) THE TRANSITION CHAINS 23.6

Assume that the dynamical system has transition tori $T_1, T_2, \ldots, T_s$. These tori will be said to form a *transition chain* if the departing whisker M_i^+ of every preceding torus T_i is transverse to the arriving whisker M_{i+1}^- of the following torus T_{i+1} at some point of their intersection (see

Figure 23.7):

$$M_1^+ \cap M_2^- \neq \emptyset,\; M_2^+ \cap M_3^- \neq \emptyset, \dots, M_{s-1}^+ \cap M_s^- \neq \emptyset .$$

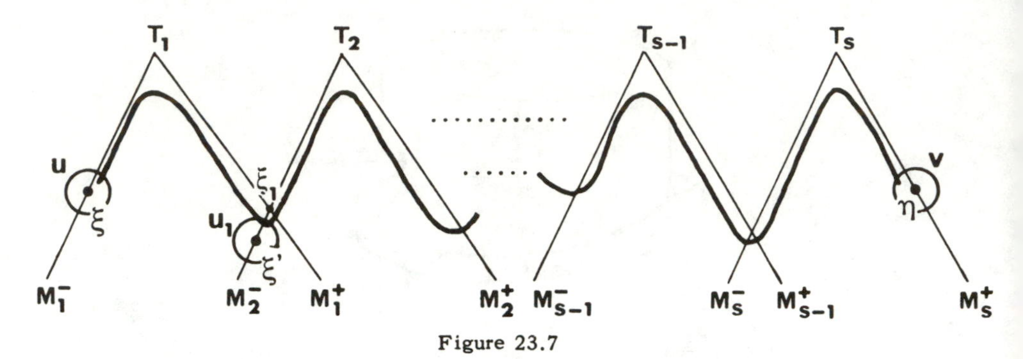

Figure 23.7

LEMMA 23.8

Let $T_1, T_2, \dots, T_s$ be a transition chain. Then, an arbitrary neighborhood U of an arbitrary point $\xi \in M_1^-$ is connected with an orbit $\zeta(t)$ to an arbitrary neighborhood V of an arbitrary point $\eta \in M_s^+$:

$$\zeta(0) \in U, \quad \zeta(t) \in V \text{ for a certain } t.$$

Proof:

Consider the future $\Omega = \bigcup_{t>0} U(t)$ of U. Since T_1 is a transition torus, then Ω obstructs M_1^+ at $\xi_1 = M_1^+ \cap M_2^-$. Thus, M_2^- intersects the open set Ω. Let ξ_1' be a point of $M_2^- \cap \Omega$, then there exists a neighborhood U_1 of ξ_1' that belongs to Ω. The future of U_1 belongs to Ω and it is sufficient to perform the same argument s times to prove that Ω obstructs M_s^+ at η. (Q. E. D.)

(D) AN UNSTABLE SYSTEM 23.9

Let $\Omega = R^2 \times T^3$ be the five-dimensional space[20] $I_1, I_2\,;\ \phi_1, \phi_2, t$ (ϕ_1, ϕ_2, t taken mod 2π). The Hamiltonian, depending on the parameters ε, μ will have the form:

---

[20] It is easy to construct a conservative system with the Hamiltonian (23.10).

$$H = \tfrac{1}{2}(I_1^2 + I_2^2) + \varepsilon \cos(\phi_1 - 1)[1 + \mu B], \quad B = \sin \phi_2 + \cos t$$

In other words, we consider the system of differential equations:

$$\text{(23.10)} \qquad \dot{\phi}_1 = I_1 ; \quad \dot{\phi}_2 = I_2, \quad \dot{I}_1 = \varepsilon \cdot \sin \phi_1 [1 + \mu B],$$

$$\dot{I}_2 = \varepsilon (1 - \cos \phi_1) \mu \cos \phi_2,$$

where $\mu \ll \varepsilon \ll 1$.

THEOREM 23.11

Assume $0 < A < B$. *For every* $\varepsilon > 0$ *there exists a* $\mu_0 = \mu_0(A, B, \varepsilon,) > 0$ *such that for* $0 < \mu < \mu_0$ *the system* (23.10) *has a solution satisfying:* $I_2(0) < A$, $I_2(t) > B$ *for a certain* t.

To prove Theorem (23.11) it is sufficient, in view of Lemma (23.8), to find a transition chain $T_1, \ldots, T_s$ such that: $I_2 < A$ on T_1 and $I_2 > B$ on T_s.

LEMMA 23.12

Each manifold T_ω *defined by the equations* $I_1 = \phi_1 = I_2 - \omega = 0$, *where* ω *is irrational, is a two-dimensional whiskered torus of the system* (23.10).

In fact:

(1) T_ω is clearly an invariant torus of (23.10);

(2) for $\mu = 0$, the three-dimensional whiskers have equations:

$$I_1 = \pm 2\sqrt{\varepsilon} \sin \frac{\phi_1}{2}, \qquad I_2 = \omega;$$

(3) for $\mu \neq 0$ and small enough, the whiskers still exist and can be found by the Hadamard method (see Chapter 3, §15). The argument of Lemma (23.5) proves that the tori T_ω are also transition tori.

Finally, making use of variation formulas of the whiskers for μ small enough,[21] the following lemma is proved:

---

[21] See Poincaré [2] and V. K. Melnikov [1].

LEMMA 23.13

Assume $A < \omega < B$. Then the departing whisker M_ω^+ of the torus T_ω intersects with the arriving whiskers $M_{\omega'}^-$, of all tori $T_{\omega'}$, which are sufficiently close (provided that $|\omega - \omega'| < K$, where $K = K(\varepsilon, \mu, A, B)$).

Proof of this lemma requires certain computations that will be found in V. Arnold [13]. These computations show also that $K \sim \mu \exp(-1/\sqrt{\varepsilon})$.

Lemmas (23.12) and (23.13) imply that the whiskered tori $T_{\omega_1}, \ldots, T_{\omega_s}$ (ω_i irrational, $|\omega_i - \omega_{i+1}| \leq K$, $\omega_1 < A$, $\omega_s > B$) form a transition chain. Application of Lemma (23.8) to this chain implies Theorem (23.11).

General References for Chapter 4

Arnold, V., Small Denominators I, *Izvestia Akad. Nauk., Math. Series* 25 1 (1961) pp. 21–86. [Transl. *Am. Math. Soc.* 46(1965) pp. 213–284.]

Small Denominators II, *Usp. Math. Nauk.* No. 5 (1963) pp. 13–40. [*Russian Math. Surveys* no. 5 (1963) pp. 9–36.]

Small Denominators III, *Usp. Math. Nauk.* No. 6(1963) pp. 89–192. [*Russian Math. Surveys* No. 6(1963) pp. 85-193.]

Birkhoff, G. D., *Dynamical Systems*, New York (1927).

Moser, J., On Invariant Curves of Area-Preserving Mappings of an Annulus, *Göttingen Nachr.* No. 1 (1962).

Poincaré, H., *Les méthodes nouvelles de la mécanique céleste*, I, II, III, Gauthier-Villars, Paris (1892, 1893, 1899).

Siegel, C. L., *Vorlesungen über Himmelsmechanik*, Springer, Berlin (1956).

APPENDIX 1

THE JACOBI THEOREM

(See Example 1.2)

Let $S^1 = \{x \pmod 1\}$ be the circle and ϕ be the translation: $x \to x+\omega \pmod 1$, $\omega \in R$. *Each orbit of* ϕ *is everywhere dense if, and only if,* ω *is irrational.*

Proof: Assume ω rational:

Let $\omega = p/q$ where $p, q \in Z$, $q > 0$. Here ϕ^q is the identity transformation. Since every point of our circle is left fixed by ϕ^q, every orbit is closed and consists of a finite set of points.

Assume ω irrational:

Let x be an arbitrary point of S^1. The $\phi^n x$'s are distinct for

$$\phi^n x = \phi^m x$$

implies that $(n-m)\omega \in Z$ and then $n = m$. Thus each orbit consists in an infinite number of distinct points. Since S^1 is compact, this orbit has a limit point. Consequently, for any $\varepsilon > 0$ there are distinct integers n and m such that:

$$|\phi^n x - \phi^m x| < \varepsilon \ .$$

Setting $|n-m| = p$ and observing that ϕ is length-preserving, we have:

$$|\phi^p x - x| < \varepsilon \ .$$

Thus, $\phi^p x, \phi^{2p} x, \ldots, \phi^{kp} x, \ldots$ partition S^1 into segments of length less than ε. Our theorem is proved, for ε is arbitrary. An N-dimensional extension of this theorem reads as follows:

Let $T^n = \mathbf{R}^n/\mathbf{Z}^n$ *be the n-dimensional torus and* ϕ *be the translation:* $\mathbf{x} \to \mathbf{x} + \omega \pmod 1$, $\omega \in \mathbf{R}^n$. *Each orbit of* ϕ *is everywhere dense if, and only if,* $\mathbf{k}\cdot\omega \in \mathbf{Z}$ *and* $\mathbf{k} \in \mathbf{Z}^n$ *imply* $\mathbf{k} = 0$.

In the continuous case we have: *Let* ϕ_t *be the translation* $\mathbf{x} \to \mathbf{x} + t\omega \pmod 1$, $t \in \mathbf{R}$, $\omega \in \mathbf{R}^n$. *Each orbit of* ϕ_t *is everywhere dense if, and only if,* $\omega\cdot\mathbf{k} = 0$ *and* $\mathbf{k} \in \mathbf{Z}^n$ imply $\mathbf{k} = \mathbf{0}$.

APPENDIX 2

GEODESIC FLOW OF THE TORUS

(See Example 1.7)

V is the two-dimensional torus, that is, the surface of revolution of a circle of radius r about a line $0z$ in the plane of the circle at the distance $1\ (> r)$ from the center of the circle. Equations of V in geographical coordinates are:

$$x = (1 + r\cos\psi)\cos\phi$$

$$y = (1 + r\cos\psi)\sin\phi$$

$$z = r\sin\psi,$$

where ϕ is the latitude and ψ is the longitude.

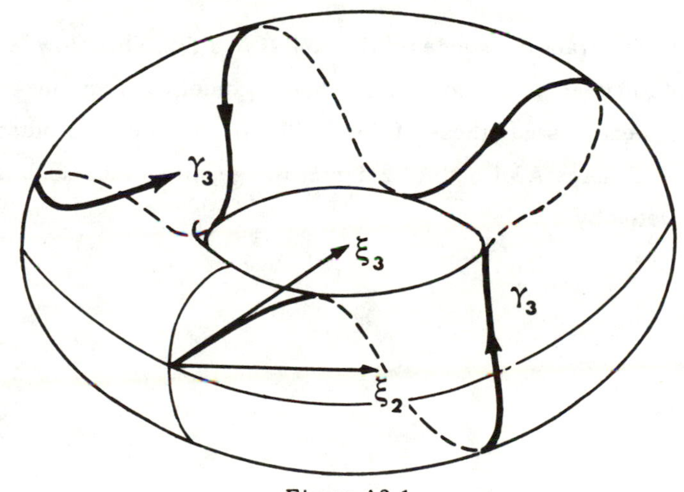

Figure A2.1

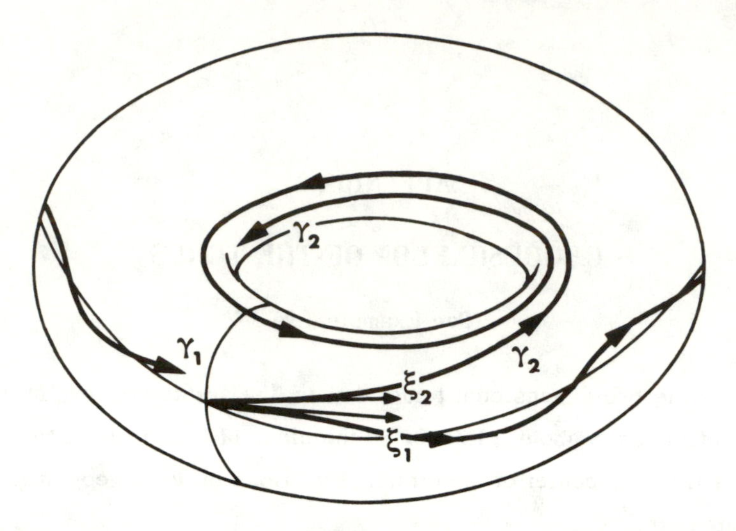

Figure A2.2

Conservation of energy and conservation of angular momentum about $0z$ give equations of the geodesics:

$$r^2 \dot{\psi}^2 + (1 + r \cos \psi)^2 \cdot \dot{\phi}^2 = h = \text{constant}.$$

$$\dot{\phi} \cdot (1 + r \cos \psi)^2 = k = \text{constant}.$$

If $h = 1$ we obtain the geodesic flow on $M = T_1 V$. This flow is invariant under rotations $\phi \to \phi + \text{constant}$. Thus, in order to study the geodesics it is sufficient to study those of them with initial point on a prescribed meridian. Figures A2.1 and A2.2 depict the generic cases γ_1 and γ_3 that are separated by γ_2.

APPENDIX 3

THE EULER-POINSOT MOTION

(See Example 1.7)

It is the motion of a heavy rigid body fixed at its center of gravity. The rigid body is a system with three degrees of freedom and a six-dimensional phase space. There exist four independent single-valued first integrals: the energy T and three components of the angular momentum vector $\mathbf{m}$. These four functions of position in the phase space do not change their values for a given motion. The points of the six-dimensional phase space for which the four functions have given values form, in general, a two-dimensional manifold M. *These manifolds* $M(T, \mathbf{m})$ *are tori*. In fact, M is invariant, and so the phase-velocity vector at each point of M is tangent to M; consequently M admits a vector field without singular point. It is evident that M is orientable and compact. The only compact two-dimensional orientable manifold admitting a tangential vector field without singular point is, as is well-known, the torus. On the other hand, M being invariant under the dynamical flow ϕ_t, admits an invariant measure μ (Liouville theorem). Thus (M, μ, ϕ_t) *is a classical system.*

A canonical transformation makes (M, μ, ϕ_t) into a system of the form: $\dot{x} = 1$, $\dot{y} = \alpha$ (Example 1.2; see Appendix 26). Thus *the Euler-Poinsot motion is,* in general, *quasi-periodic* and the orbits are dense on M. Its two periodic components are called, respectively, precession and rotation (Euler [1]).

APPENDIX 4

GEODESIC FLOWS OF LIE GROUPS

(See Example 1.7)

The geodesic flow of a Lie group carrying a left-invariant (or right-invariant) metric has important applications:

The geodesic flow of the connected component $SO(3)$ of the group of the rotations of the three-dimensional Euclidean space corresponds to the rotations of a heavy solid moving around a fixed point. Each orbit corresponds to a motion.

The homotheties of positive ratio and the translations of the n-dimensional space $\mathbf{R}^n$ form a group that generates the geodesic flow of the $(n+1)$-dimensional space of constant negative curvature.

Then, let us consider the group $S\,\mathrm{Diff}(\mathfrak{D})$ of the measure-preserving diffeomorphisms of a compact Riemannian domain $\mathfrak{D}$. The corresponding algebra consists in divergence-free vector fields V on $\mathfrak{D}$: $\operatorname{div} V = 0$. The energy $< V, V > = \int_{\mathfrak{D}} V^2\, dx$ is a positive definite quadratic form on the space of such vector fields, and defines some right-invariant Riemannian metric on the group $S\,\mathrm{Diff}(\mathfrak{D})$.

The geodesics of this metric are just flows of perfect (incompressible and unviscous) fluid in $\mathfrak{D}$. The Riemannian curvature of this infinite dimensional "manifold" $S\,\mathrm{Diff}(\mathfrak{D})$ can be computed. For example, if $\mathfrak{D} = T^2$ is the torus $\{x, y \bmod 2\pi\}$ with its usual metric, then the sectional curvature is nonpositive in every section containing the laminar flow:

$$V_x = \cos y, \quad V_y = 0.$$

See also V. I. Arnold [1] and L. Auslander, L. Green, F. Hahn [1].

APPENDIX 5

THE PENDULUM

(See 1.13)

The equation of the pendulum is: $\ddot{q} + k \sin q = 0$, where k is a positive constant; it is equivalent to the system:

$$\begin{cases} \dot{q} = p \\ \dot{p} = -k \sin q . \end{cases}$$

The Hamiltonian function is $H = (p^2/2) - k \cdot \cos q$ and Figure (A5.1) depicts the orbits. The system is invariant under the symmetry $p \to -p$ and the translations:

$$(q, p) \to (q + 2K\pi, p), \quad K \in \mathbf{Z} .$$

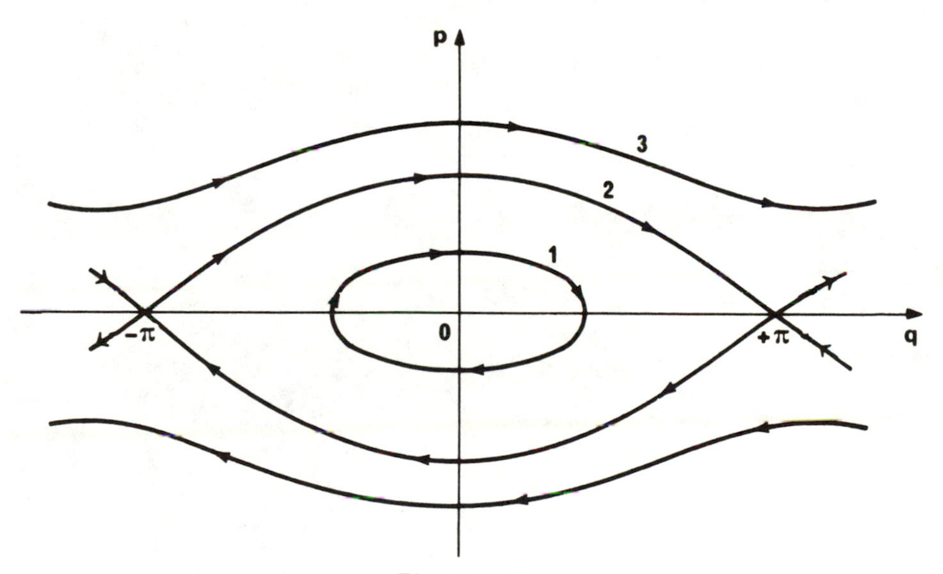

Figure A5.1

The points $(k\pi, 0)$ are critical points: points $(2\pi k, 0)$ are centers (stable equilibrium) and points $((2k+1)\pi, 0)$ are saddle points (unstable equilibrium). The orbits split into three types: the orbits 1 (small oscillations); the separatrices 2 joining two saddle points; and the orbits 3 (complete rotation around the hanging point).

The natural phase space is no longer the plane (p, q) but it is rather the cylinder $(q \bmod \pi, p)$. This is an example of global Hamiltonian flow.

APPENDIX 6

MEASURE SPACE

(See Chapter 1, Section 2)

An σ-algebra $\mathcal{B}$ defined on M is a class of subsets of M that is closed under the formation of complement and countable union. A nonnegative (possibly infinite) countably additive set function μ defined on $\mathcal{B}$ is called a measure $(M, \mathcal{B}, \mu)$ is called a measure space; $\mathcal{B}$ is the family of the measurable subsets of M and μ is the measure. (See Halmos [2] and Rohlin [3] for these concepts.)

In fact, the object of interest is not $(M, \mathcal{B}, \mu)$ but equivalence classes (M, μ) of measure spaces. Let us make this point clearer. Let A and B be elements of $\mathcal{B}$, we set $A = B \pmod 0$ if $\mu(A \cup B - A \cap B) = 0$. This relation is an equivalence relation and the equivalence class of the empty set consists in the sets of measure zero. We denote the quotient of $\mathcal{B}$ under this equivalence relation by $\mathcal{B} \pmod 0$; it is a Boolean algebra, for the following properties are readily proved:

$$A_1, A_2, B_1, B_2 \in \mathcal{B}, \quad A_1 = A_2 \pmod 0, \quad B_1 = B_2 \pmod 0$$

imply

$$A_1 \cup B_1 = A_2 \cup B_2 \pmod 0, \; A_1 \cap B_1 = A_2 \cap B_2 \pmod 0,$$
$$M - A_1 = M - A_2 \pmod 0 .$$

If $A = B \pmod 0$, then $\mu(A) = \mu(B)$ and μ can be regarded as a function defined on $\mathcal{B} \pmod 0$.

To study abstract dynamical systems one neglects sets of measure zero. This means that one replaces the study of $(M, \mathcal{B}, \mu)$ by the study of

$(M, \mathcal{B} \text{ (mod 0)}, \mu)$, which we still denote unambiguously by (M, μ), because the function μ defines completely $\mathcal{B}$ (mod 0) (but not $\mathcal{B}$). In other words, we identify $(M, \mathcal{B}, \mu)$ and $(M, \mathcal{B}', \mu')$ if their measurable σ-algebras $\mathcal{B}$ (mod 0) and $\mathcal{B}'$ (mod 0) coincide.

Let (M, μ) and (M', μ') be two measure spaces. A mapping ϕ: $M \to M'$ is called an *homomorphism modulo zero* if:

(a) ϕ is defined on $M - I$, where I is a set of measure zero (possibly empty);

(b) $I' = M' - \phi(M-I)$ has measure zero, that is ϕ is onto, up to a subset I' of M' of measure zero;

(c) ϕ is measure-preserving, that is each equivalence class of $\mathcal{B}'$ (mod 0) contains a representative, say A', such that $\phi^{-1}(A')$ exists and belongs to $\mathcal{B}$ with:

$$\mu[\phi^{-1}(A)] = \mu'[A'].$$

Thus, ϕ induces an homomorphism of Boolean measurable algebras: ϕ^{-1}: $\mathcal{B}'$ (mod 0) $\to \mathcal{B}$ (mod 0).

If $(M, \mu) = (M', \mu')$ then ϕ is called an *endomorphism (mod 0)*. If both of the mappings ϕ and ϕ^{-1} are homomorphisms (mod 0), then ϕ is called an *isomorphism (mod 0)*; if, additionally, (M, μ) and (M', μ') coincide, then ϕ is called an *automorphism (mod 0)*.

APPENDIX 7

ISOMORPHISM OF THE BAKER'S TRANSFORMATION AND $B(\frac{1}{2}, \frac{1}{2})$

(See Example 4.5)

We need to construct an isomorphism (mod 0) f making the following diagram commutative:

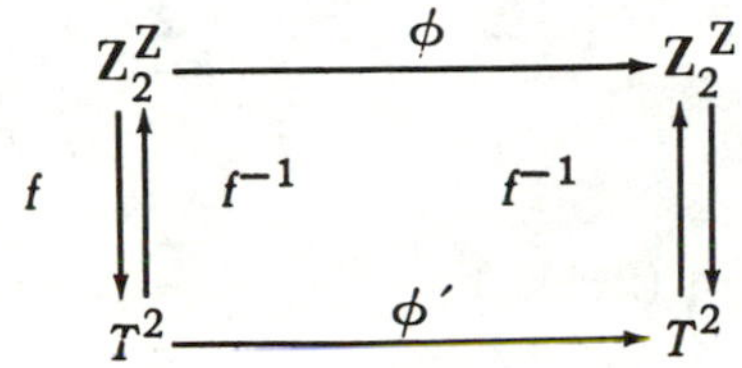

$$\begin{array}{ccc} \mathbf{Z}_2^{\mathbf{Z}} & \xrightarrow{\phi} & \mathbf{Z}_2^{\mathbf{Z}} \\ f \downarrow \uparrow f^{-1} & & f^{-1} \uparrow \downarrow f \\ T^2 & \xrightarrow{\phi'} & T^2 \end{array}$$

Definition of f. Let $m = \ldots, a_{-1}, a_0, a_1, \ldots$ be a point of $\mathbf{Z}_2^{\mathbf{Z}}$. We set $f(m) = (x, y)$, where

$$x = \sum_{i=0}^{\infty} \frac{a_{-i}}{2^{i+1}}, \qquad y = \sum_{i=1}^{\infty} \frac{a_i}{2^i}. \tag{A 7.1}$$

The mapping f is a bijection, except on the elements (x, y) of T^2 for which x or y is a dyadic fraction. Such elements are denumerable and so constitute a set of measure zero.

f is measure-preserving. It is sufficient to prove it for a generator $A_i^j = \{m \mid a_i = j\}$ of the measure algebra of $\mathbf{Z}_2^{\mathbf{Z}}$: the set

$$f(A_i^j) = \left\{ \left(\sum_{k=0}^{\infty} \frac{a_{-k}}{2^{k+1}}, \sum_{k=1}^{\infty} \frac{a_k}{2^k} \right) \middle| \; a_i = j \right\}$$

consists in $2^{|i|}$ rectangles, the sides of which are 1 and $1/2^{|i|+1}$. Thus, we get:

$$\mu'[f(A_i^j)] = \frac{1}{2} = \mu(A_i^j) \; .$$

The diagram is commutative. Let x and y be given by formula (A 7.1), we have:

$$f^{-1}(x, y) = \ldots, a_{-1}, a_0, a_1, \ldots$$

$$\phi \cdot f^{-1}(x, y) = \ldots, a'_{-1}, a'_0, a'_1, \ldots,$$

where $a'_i = a_{i-1}$,

$$f\phi f^{-1}(x, y) = \left(\sum_{k=1}^{\infty} \frac{a_{-k}}{2^k} \, , \; \sum_{k=0}^{\infty} \frac{a_k}{2^{k+1}} \right) ;$$

that is to say:

$$f\phi f^{-1}(x, y) = \begin{cases} (2x, \tfrac{1}{2}y) & \text{if } a_0 = 0, \text{ i.e., } 0 \leq x < \tfrac{1}{2} \\ (2x, \tfrac{1}{2}(y+1)) & \text{if } a_0 = 1, \text{ i.e., } \tfrac{1}{2} \leq x < 1 . \end{cases}$$

Consequently: $f\phi f^{-1} = \phi'$. (Q. E. D)

APPENDIX 8

LACK OF COINCIDENCE EVERYWHERE OF SPACE MEAN AND TIME MEAN

(See Remark 6.5)

Consider again the dynamical system of Example (1.16): M is the torus $\{(x, y) \bmod 1\}$, the measure is $dx\,dy$ and the automorphism ϕ is:

$$\phi(x, y) = (x + y,\ x + 2y) \pmod 1 .$$

This automorphism ϕ induces a linear mapping $\tilde{\phi}$ of the covering plane $\tilde{M} = \{(x, y)\}$. The matrix

$$\begin{pmatrix} 1 & 1 \\ 1 & 2 \end{pmatrix}$$

of $\tilde{\phi}$ has two proper values: $0 < \lambda_2 < 1 < \lambda_1$. The line

$$\begin{cases} x = s \\ y = (\lambda_2 - 1)s, \quad s \in \mathbf{R} \end{cases}$$

projects onto a curve γ of M under the natural projection $\tilde{M} \to M$. This curve γ is invariant under ϕ and γ is dense on M, for $\lambda_2 - 1$ is irrational (Jacobi's theorem, Appendix 1). Let $m = (x, y)$ be a point of γ. Of course, we have:

$$\phi^n(m) = (\lambda_2^n x, \lambda_2^n y) \pmod 1$$

and $0 < \lambda_2 < 1$ implies:

$$\lim_{n=\infty} \phi^n(m) = (0, 0) .$$

Consider the analytic function $f(x, y) = e^{2\pi i x}$. We have

$$\frac{1}{N}\sum_{n=0}^{N-1} f(\phi^n m) = \frac{1}{N}\sum_{n=0}^{N-1} e^{2\pi i x \lambda_2^n}.$$

Usual convergence implies Cesaro convergence and $\lim_{n=\infty} x\cdot\lambda_2^n = 0$, therefore:

$$\overset{*}{f}(m) = \lim_{N\to+\infty}\frac{1}{N}\sum_{n=0}^{N-1} f(\phi^n m) = 1 .$$

On the other hand:

$$\overline{f} = \int_M f(x, y)\,dx\,dy = \int_0^1 e^{2\pi i x}\cdot dx = 0.$$

Thus, whatever the point m of the dense subset γ be, we have: $\overset{*}{f}(m) \neq \overline{f}$, though f is analytic and ϕ is classical.

APPENDIX 9

THE THEOREM OF EQUIPARTITION MODULO 1

(See 6.6)

We prove here[1] the Theorem of Equipartition Modulo 1 due to Bohl, Sierpinskii, and Weyl: *If ϕ is a rotation of the circle M through an angle incommensurate with 2π:*

$$M = \{z \in \mathbf{C},\ |z| = 1\},\ \phi(z) = \theta \cdot z,\ \theta = e^{2\pi i\omega},\ \omega \text{ is irrational}$$

and f is a Riemannian integrable function, then the time mean of f exists everywhere and coincides with the space mean.

Proof.

1st case: $f(z) = z^p,\ p \in \mathbf{Z}$. We get:

$$\frac{1}{N}\sum_{n=0}^{N-1} f(\phi^n z) = \frac{1}{N}\sum_{n=0}^{N-1} (\theta^n z)^p = \begin{cases} 1 & \text{if } p = 0 \\ \dfrac{1}{N} \cdot z^p \cdot \dfrac{\theta^{Np} - 1}{\theta^p - 1} & \text{if } p \neq 0\,. \end{cases}$$

Since ω is irrational we have $\theta^p - 1 \neq 0$ and $|\theta^{pN} - 1| < 2$, so we get:

$$\overline{f} = \overset{*}{f}(z) = \begin{cases} 1 & \text{if } p = 0 \\ 0 & \text{if } p \neq 0\,. \end{cases}$$

2nd case: f is a trigonometrical polynomial, that is

$$f(z) = \Sigma\, a_p z^p, \quad p \in \mathbf{Z}, \quad z \in M,$$

---

[1] Compare to G. Polya and G. Szego [1] p. 73.

in which $a_p = 0$, except for a finite number of them. From the first case one deduces at once:

$$\overset{*}{f}(z) = a_0 = \overline{f} \; .$$

3rd case: f is real-valued and Riemannian-integrable. To every $\varepsilon > 0$ correspond two trigonometrical polynomials P_ε^- and P_ε^+ such that:

$$P_\varepsilon^-(z) < f(z) < P_\varepsilon^+(z) \text{ for every } z \in M$$

and

$$\int_M (P_\varepsilon^+(z) - P_\varepsilon^-(z)) d\mu < \varepsilon \; .$$

From the second case we deduce:

$$\text{(A 9.1)} \qquad \int_m P_\varepsilon^- \cdot d\mu \leq \liminf_{N\to\infty} \frac{1}{N} \sum_0^{N-1} f(\phi^n z)$$

$$\leq \limsup_{N\to\infty} \frac{1}{N} \sum_0^{N-1} f(\phi^n z) \leq \int_M P_\varepsilon^+ \cdot d\mu \; .$$

Consequently, lim sup – lim inf < ε. Since ε is arbitrary we have lim sup = lim inf = lim = $\overset{*}{f}(z)$, which exists everywhere. Relation (A9.1) implies that $\overset{*}{f}(z)$ is constant, whence:

$$\overset{*}{f}(z) = \overline{f} \; . \qquad \text{(Q. E. D.)}$$

Extension to translations of the torus T^n is obvious: the time mean and the space mean of a Riemannian-integrable function coincide everywhere if, and only if, the orbits are everywhere dense.

APPENDIX 10

SOME APPLICATIONS OF ERGODIC THEORY TO DIFFERENTIAL GEOMETRY

The Birkhoff theorem was used by A. Avez [1] to prove the following:

Let V be a compact n-dimensional Riemannian manifold without conjugate point, then the proper values of the operator

$$\Delta - \frac{R}{n-1}$$

are nonnegative (Δ is the Laplacian $-\nabla^{\alpha}\nabla_{\alpha}$, R is the scalar curvature). In particular (L. W. Green):

$$\int_M R \cdot \eta \leq 0 .$$

APPENDIX 11

ERGODIC TRANSLATIONS OF TORI

(See Example 7.8)

We prove that translations of tori (Examples 1.2 and 1.15) are ergodic if and only if, their orbits are everywhere dense (or if, and only if, the time mean and the space mean of a continuous function coincide everywhere).

Let M be the n-dimensional torus $\{e^{2\pi i\mathbf{x}} | \mathbf{x} \in \mathbf{R}^n\}$, where $\mathbf{x} = (x_1, \ldots, x_n)$ and $e^{2\pi i\mathbf{x}}$ means $(e^{2\pi i x_1}, \ldots, e^{2\pi i x_n})$. The measure of M is the usual product measure μ. The translation is:

$$\phi: \ e^{2\pi i\mathbf{x}} \longrightarrow e^{2\pi i(\mathbf{x}+\omega)}, \qquad \omega \in \mathbf{R}^n .$$

THEOREM.

(M, μ, ϕ) is ergodic if, and only if $\mathbf{k}\cdot\omega \in \mathbf{Z}$ *and* $\mathbf{k} \in \mathbf{Z}^n$ *imply* $\mathbf{k} = 0$.

Proof:

Let f be a measurable invariant function. Its Fourier coefficients are:

$$a_k = \int_M e^{-2\pi i\mathbf{k}\cdot\mathbf{x}} \cdot f(\mathbf{x})d\mu .$$

The Fourier coefficients of $f(\phi\mathbf{x})$ are:

$$b_{\mathbf{k}} = \int_M e^{-2\pi i\mathbf{k}(\mathbf{x}-\omega)} \cdot f(\mathbf{x})d\mu = e^{2\pi i\mathbf{k}\cdot\omega} \cdot a_{\mathbf{k}} .$$

The invariance of f is equivalent to $b_{\mathbf{k}} = a_{\mathbf{k}}$ for any $\mathbf{k}$, that is $a_{\mathbf{k}} = 0$ or $\mathbf{k}\cdot\omega \in \mathbf{Z}$.

If the ω_i's are integrally independent, the second case occurs only for $\mathbf{k} = 0$. Thus a_0 is the only Fourier coefficient possibly nonzero, f is constant, and (M, μ, ϕ) is ergodic (see 7.2).

If a $\mathbf{k} \neq 0$ exists such that $\mathbf{k} \cdot \omega \in \mathbf{Z}$, then $f(\mathbf{x}) = e^{2\pi i \mathbf{k} \cdot \mathbf{x}}$ is a non-constant invariant function and (M, μ, ϕ) is not ergodic.

Remark.

In the continuous case (M, μ, ϕ_t), where

$$\phi_t\colon \; e^{2\pi i \mathbf{x}} \to e^{2\pi i(\mathbf{x} + t\omega)},$$

we have a similar result: (M, μ, ϕ_t) is ergodic, if, and only if, $\mathbf{k} \in \mathbf{Z}^n$ and $\mathbf{k} \cdot \omega = 0$ imply $\mathbf{k} = 0$ (or if, and only if, the orbits are everywhere dense; see Jacobi's theorem).

APPENDIX 12

THE TIME MEAN OF SOJOURN

(See Chapter 2, Section 7)

THEOREM A12.1

An abstract dynamical system (M, μ, ϕ_t) is ergodic if, and only if, the sojourn time $\tau(T)$ in an arbitrary measurable set A of an orbit

$$\{\phi_t x \mid 0 \leq t \leq T\}$$

is asymptotically proportional to the measure of A:

$$\tau(T) = \text{measure}\,\{t \mid 0 \leq t \leq T,\ \ \phi_t x \in A\}$$

$$\lim_{T\to\infty} \frac{\tau(T)}{T} = \mu(A), \tag{A12.2}$$

for all measurable A and almost every initial point $x \in M$.

Proof:

Assume (M, μ, ϕ_t) is ergodic and A is measurable. We have $\overset{*}{f}(x) = \bar{f}$ for every $f \in L_1(M, \mu)$ and for almost every $x \in M$ (see 7.1). Take $f = \mathcal{X}_A$ (characteristic function of the set A), we obtain:

$$\lim_{T\to\infty} \frac{\tau(T)}{T} = \lim_{T\to\infty} \frac{1}{T} \int_0^T \mathcal{X}_A(\phi_t x)dt = \int_M \mathcal{X}_A(x)d\mu = \mu(A)$$

for almost every x.

The converse is derived at once: (A12.2) implies ergodicity. It is sufficient to observe that the characteristic functions $\mathcal{X}_A$ generate $L_1(M, \mu)$. Theorem (A12.1) clearly holds in the discrete case (M, μ, ϕ).

EXAMPLES A12.3: TRANSLATIONS OF TORI

Let M be the n-dimensional torus $\{e^{2\pi i x} \mid x \in R^n\}$, μ the usual measure, and ϕ the translation:

$$\phi: e^{2\pi i x} \longrightarrow e^{2\pi i(x+\omega)}, \qquad \omega \in R^n .$$

If $k \in Z^n$ and $k \cdot \omega \in Z$ imply $k = 0$, then (M, μ, ϕ) is ergodic (see Appendix 11). Thus, relation (A12.2) holds for almost every initial point. This can be rephrased as follows: denote by $\tau(N, A)$ the number of elements of the sequence

$$e^{2\pi i x}, \quad e^{2\pi i(x+\omega)}, \ldots, \quad e^{2\pi i(x+(N-1)\omega)}$$

that belong to A, then:

$$\lim_{N \to \infty} \frac{\tau(N, A)}{N} = \mu(A) \tag{A12.4}$$

for almost every initial point $e^{2\pi i x}$. If A is Jordan-measurable, that is, if $\mathcal{X}_A$ is Riemannian-integrable, then (A12.4) holds for every initial point. To prove it, it is sufficient to use the theorem of Appendix 9 and to take $f = \mathcal{X}_A$. Extension to the continuous case holds good. This result is known as the theorem of equipartition modulo 1[1] and is due to P. Bohl [1], W. Sierpinskii, and H. Weyl [1], [2], [3]. It is one of the first ergodic theorems. Historically, it originated from an attempt to solve the Lagrange problem of the mean motion of the perihelion (see Example 3.1 and Appendix 13).

Here follow some applications.[2]

APPLICATION A12.5: DISTRIBUTION OF THE FIRST DIGITS OF 2^n (see Example 3.2)

The first digit of 2^n is equal to k if, and only if:

$$k \cdot 10^r \leq 2^n < (k+1) \cdot 10^r ,$$

---

[1] F. P. Callahan [1] gave an elementary proof.

[2] The reader will find further applications to various fields in: *Compositio Mathematica, V 16,* fascicles 1, 2.

that is to say:

$$r + \mathrm{Log}_{10}k \leq n\,\mathrm{Log}_{10}2 \leq r + \mathrm{Log}_{10}(k+1) .$$

Set $\alpha = \mathrm{Log}_{10}2$ and $(n \cdot \alpha) = n\alpha - [n \cdot \alpha]$, where [ ] means the integer part. The above inequality may be written:

$$\mathrm{Log}_{10}k \leq (n\alpha) < \mathrm{Log}_{10}(k+1) .$$

Now, we turn to the dynamical system consisting of the one-dimensional torus $M = \{e^{2\pi i x} \mid x \in \mathbf{R}\}$, the usual measure μ, and the translation ϕ: $e^{2\pi i x} \to e^{2\pi i(x+\alpha)}$. (M, μ, ϕ) is ergodic, for α is irrational (see Example 7.8). Thus, the sequence $\{(n\alpha) \mid n \in \mathbf{N}\}$ is equidistributed. In particular, take $A = [\mathrm{Log}_{10}k,\ \mathrm{Log}_{10}(k+1)]$ in relation (A12.4), we have:

$$\lim_{N \to \infty} \frac{\tau(N, A)}{N} = \mu(A) = \mathrm{Log}_{10}(1 + \frac{1}{k}) .$$

But $\tau(N, A)$ is nothing but the number of elements of the sequence 1, 2,..., 2^{N-1}, the first digit of which is k. Thus, if we go back to the notation of Example (3.2), we have:

$$p_7 = \mathrm{Log}_{10}(1 + \frac{1}{7}) .$$

Consequently, the proportion of 7's is greater than the proportion of 8's in the sequence of the first digits of $\{2^n \mid n = 1, 2, \ldots\}$. This is not what one expects from an inspection of the first terms: 1, 2, 4, 8, 1, 3, 6, 1, 2, 5, ... This is due to the fact that $\alpha = 0,30103\cdots$ is very close to 3/10.

REMARK A12.6

Since the sojourn time in a domain A of a point of an ergodic system is proportional to the measure of A, it is natural to ask about the dispersion. Let us mention some results due to Sinai [1]. Let ϕ_t be the geodesic flow of the unitary tangent bundle T_1V of a surface V of constant negative curvature. If A is a domain of T_1V with piecewise differentiable boundary, then the mean sojourn time of a geodesic $\phi_t x$ in this domain has a Gaussian distribution and verifies the central limit theorem:

$$\lim_{T\to\infty} \mu \left\{ x \;\middle|\; \frac{\tau_T(x)}{T} - \mu(A) < \frac{a}{\sqrt{T}} \right\} = \frac{1}{\sqrt{2\pi}} \int_{-\infty}^{Ca} e^{-u^2/2} \cdot du$$

where $\tau_T(x)$ = measure $\{ t \mid \phi_t x \in A,\ 0 \le t \le T\}$ and C is a constant.

APPENDIX 13

THE MEAN MOTION OF THE PERIHELION

(See Example 3.1 and Appendix 12)

The problem of mean motion arises from the theory of the secular perturbations of the planetary orbits (Lagrange [1]). One asks for the existence and estimate of:

(A13.1) $$\Omega = \lim_{t\to+\infty} \frac{1}{t} \operatorname{Arg} \sum_{k=1}^{n} a_k \cdot e^{i\omega_k t} ,$$

where ω_k, $t \in \mathbf{R}$ and $a_k \neq 0$, $a_k \in \mathbf{C}$.

In other words, if one considers a plane linkage $A_0 A_1 \cdots A_n$ consisting of the links $A_{k-1} A_k$ of fixed lengths $|a_k|$, moving with constant rotation-velocity ω_k, we are interested in the mean rotation-velocity of the vector $A_0 A_n$ (see Figure A13.1).

THEOREM A13.2 (See H. Weyl, [1]–[5].)

Assume that the ω_k's are integrally independent, that is:

(A13.3) $$\omega \cdot \mathbf{k} = 0 \text{ and } \mathbf{k} \in \mathbf{Z}^n \text{ imply } \mathbf{k} = 0.$$

Then the mean motion Ω exists and is expressed as:

(A13.4) $$\Omega = p_1\omega_1 + \cdots + p_n\omega_n, \quad p_k \geq 0, \quad \Sigma p_k = 1 .$$

The p_k's depend on the $|a_k|$ only. If[1] $p(a_k;\, a_1, \ldots, \hat{a}_k, \ldots, a_n)$ *is the probability that an $(n-1)$-linkage with prescribed sides $a_1, \ldots, \hat{a}_k, \ldots, a_n$ spans a distance inferior to a_k* (see formula A13.12), *then:*

---

[1] $\hat{}$ means cancellation.

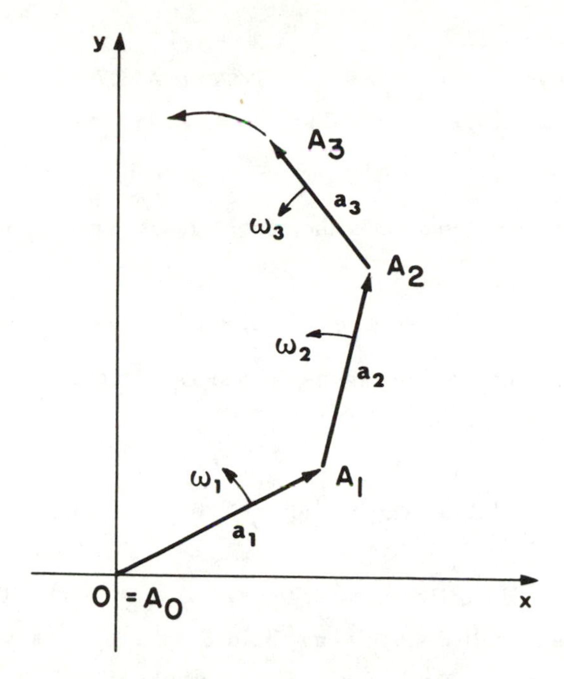

Figure A13.1. Case n = 3: initial position of the linkage.

(A13.5) $$p_k = p(|a_k|\,;\ |a_1|, \ldots, |\hat{a}_k|, \ldots, |a_n|)\ .$$

In particular, for $n = 3$, if there exists a triangle the sides of which are $|a_1|$, $|a_2|$, $|a_3|$ and the angles of which are A_1, A_2, A_3, then (Bohl's formula):

(A13.6) $$\Omega = \frac{A_1\omega_1 + A_2\omega_2 + A_3\omega_3}{\pi}\ .$$

The case in which no triangle can be constructed was investigated by Lagrange [1]. In the general case A. Wintner [1] found the expression:

$$p(a_1\,;\ a_2, \ldots, a_n) = a_1 \int_0^\infty J_1(a_1\rho) \prod_{k=2}^{n} J_0(a_k\rho)\, d\rho$$

in terms of the Bessel functions J_0 and J_1. Relation $\Sigma\, p_k = 1$ provides an "addition" theorem for these functions.

Proof of Theorem A13.2

THE CORRESPONDING DYNAMICAL SYSTEM A13.7

Let us consider the dynamical system (M, μ, ϕ_t), where

$$M = T^n = \{z \mid z = (z_1, \ldots, z_n)\}, \quad z_k = e^{2\pi i\theta_k}, \quad \theta_k \in \mathbf{R},$$

is the n-dimensional torus, μ is the usual measure, and ϕ_t is the translation group:

$$\phi_t z = (z_1 e^{i\omega_1 t}, \ldots, z_n e^{i\omega_n t}) .$$

The phase space of the n-linkage is M, and ϕ_t depicts the movement. Let us define a function α on M by:

$$\text{(A13.8)} \qquad \alpha(z) = \operatorname{Arg} \sum_{k=1}^{n} |a_k| z_k, \quad 0 \leq \alpha < 2\pi .$$

This function is discontinuous over the slit $\Sigma = \{z \mid \alpha(z) = 0\}$, and is not defined on the so-called singular manifold $S = \{z \mid \Sigma_1^n |a_k| z_k = 0\}$ which consists of all possible states of a closed n-linkage with the prescribed sides $|a_k|$. Nevertheless, the function:

$$\text{(A13.9)} \qquad \beta(z) = \frac{d}{dt} \alpha(\phi_t z)\Big|_{t=0}$$

is analytic outside of S. The limit (A13.1), if it exists, is nothing but the time mean $\overset{*}{\beta}$ of β:

$$\text{(A13.10)} \qquad \left| \operatorname{Arg} \sum_{k=1}^{n} a_k \cdot e^{i\omega_k t} \right|_{t=0}^{t=T} = \int_0^T \beta(\phi_t z)dt ,$$

where

$$z = (z_1, \ldots, z_n), \quad z_k = \frac{a_k}{|a_k|} .$$

THE SPACE MEAN A13.11

The system (M, μ, ϕ_t) is ergodic, for the ω_k's are integrally independent (Appendix 11). If the function β were Riemannian-integrable, then, according to the theorem of Equipartition modulo 1 (Appendix 9), the time mean $\overset{*}{\beta} = \Omega$ would be equal to the space mean $\bar{\beta}$.

We only know (see A. Wintner [1]) that β is Lebesgue-integrable. Thus, the Birkhoff theorem implies that $\Omega = \bar{\beta}$ for almost every initial phase. This suggests the study of the space mean $\bar{\beta}$. Relation (A13.9) shows that β depends linearly on the ω_k's. Therefore, $\bar{\beta}$ depends linearly on the ω_k's:

$$\bar{\beta}(\omega) = p_1\omega_1 + \cdots + p_n\omega_n \, .$$

To compute p_1 (for instance) we set:

$$\omega_1 = 2\pi, \ \omega_2 = \cdots = \omega_n = 0 \, .$$

We have:

$$p_1 = \frac{1}{2\pi}\bar{\beta}(2\pi, 0, \ldots, 0) \, ,$$

where

$$\bar{\beta}(2\pi, 0, \ldots, 0) = \int \cdots \int_{T^{n-1}} \left(\int_0^1 \frac{\partial a}{\partial \theta_1} \, d\theta_1 \right) d\theta_2 \cdots d\theta_n \, .$$

Relation (A13.8) allows one to carry out the integration over θ_1:

$$\int_0^1 \frac{\partial a}{\partial \theta_1} \, d\theta_1 = \begin{cases} 2\pi & \text{if } \left| |a_2| e^{2\pi i\theta_2} + \cdots + |a_n| e^{2\pi i \theta_n} \right| < |a_1| \\ 0 & \text{if } \left| |a_2| e^{2\pi i\theta_2} + \cdots + |a_n| e^{2\pi i \theta_n} \right| > |a_1| \, . \end{cases}$$

Thus we obtain:

$$p_1 = p(|a_1|; \ |a_2|, \ldots, |a_n|) \, ,$$

where

(A13.12) $p_1(a_1; \ a_2, \ldots, a_n) = \text{measure}\{z \in T^n \,|\, |a_2 z_2 + \cdots + a_n z_n| < |a_1|\} \, .$

This proves relation (A13.5).

Relation $\Sigma \, p_k = 1$ is derived easily by setting

$$\omega_1 = \cdots = \omega_n = 2\pi \, .$$

EXISTENCE OF THE TIME MEAN A13.13

Hence, formulas (A13.4) and (A13.5) are proved for almost every initial phase Arg a_k. To prove them for all initial phases we use a special device inaugurated by Bohl [1] for $n = 3$ and improved by Weyl [4], [5] for $n > 3$. We define a function on the torus M by:

$N(z)$ = *algebraic number of the points of intersection of the curve* $\{\phi_t z, \ -2\pi < t < 0\}$ *with the slit* Σ.

We count +1 a point of intersection z_j for which $\beta(z_j) > 0$ and −1 if $\beta(z_j) < 0$. (See Figure A13.15.) It can be proved that $N(z)$ is bounded.

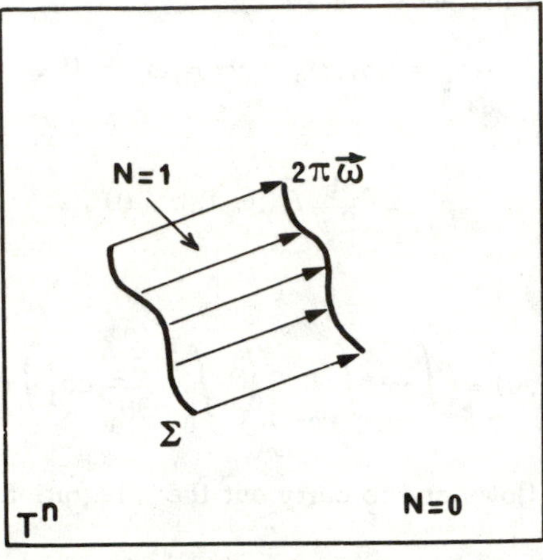

Figure A13.15

Thus, according to (A13.10), the following relation holds uniformly over T^n:

$$\text{(A13.14)} \qquad \left| \int_0^T \beta(\phi_t z)dt - \int_0^T N(\phi_t z)dt \right| < C \ .$$

Since the function N is piecewise continuous and, in particular, Riemannian-integrable, the time mean $\overset{*}{N}$ exists everywhere and is equal to $\overline{N}$ (Appendix 9). From (A13.14) one deduces that $\overset{*}{\beta} = \overset{*}{N} = \overline{N}$ exists everywhere and is constant. (Q. E. D.)

APPENDIX 14

EXAMPLE OF A MIXING ENDOMORPHISM

Let us consider the transformation:[1]

$$\phi: (x, y) \to (2x, 2y) \pmod 1$$

of the torus $M = \{(x, y) \bmod 1\}$ carrying the usual measure $dx\,dy$.

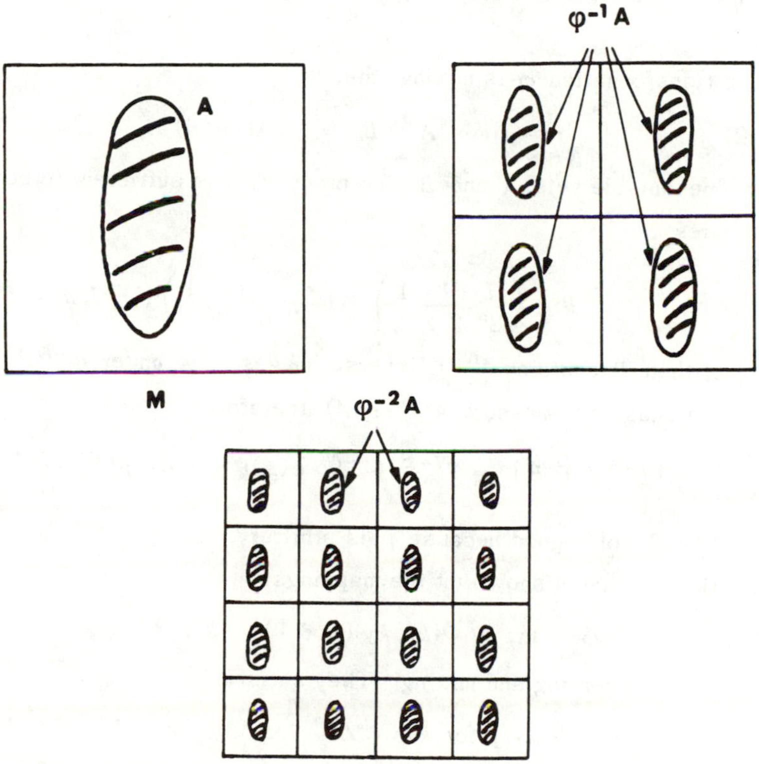

Figure A14.1

[1] Which is called "multiplication of loaves" since Figure (A14.1) shows the solution of a well-known historical problem.

To be more explicit, we write:

$$\phi(x,y) = \begin{cases} (2x, 2y) & \text{if } 0 \le x, \quad y < \tfrac{1}{2} \\ (2x, 2y-1) & \text{if } 0 \le x < \tfrac{1}{2}, \ \tfrac{1}{2} \le y < 1 \\ (2x-1, 2y) & \text{if } \tfrac{1}{2} \le x < 1, \quad 0 \le y < \tfrac{1}{2} \\ (2x-1, 2y-1) & \text{if } \tfrac{1}{2} \le x, y < 1. \end{cases}$$

The application ϕ is not one-to-one, in fact it is everywhere four-to-one. If E is a square with dyadically rational vertices, then $\phi^{-1}E$ is the union of four similar squares (see Figure). Thus, in such a case, $\mu(\phi^{-1}E) = \mu(E)$ and from there it follows easily that ϕ is measure-preserving. We have here an example of a measure-preserving mapping that is not invertible.

The transformation ϕ is mixing, that is:

$$\text{(A14.2)} \qquad \lim_{N\to\infty} \mu[\phi^{-N}A \cap B] = \mu(A)\cdot\mu(B)$$

for any measurable sets A and B. To prove it, it is sufficient to consider the squares B:

$$B = \left(\frac{l}{2^p}, \frac{l+1}{2^p}\right) \times \left(\frac{m}{2^p}, \frac{m+1}{2^p}\right), \qquad l, m \in \mathbf{Z}^+ .$$

If $N \geq p$, then B contains 4^{N-p} inverse images of A under ϕ^{-N}. Such an inverse image has measure $4^{-N}\cdot\mu(A)$, therefore:

$$\mu[\phi^{-N}A \cap B] = 4^{N-p}.(4^{-N}\cdot\mu(A)) = \mu(A)\cdot\mu(B) .$$

Thus (A14.2) holds good because p is arbitrary.

Similar arguments show that the mappings:

$$\phi_k: (x, y) \to (kx, ky)(\text{mod } 1), \qquad k \in \mathbf{Z}^+$$

are measure-preserving and mixing. They satisfy:

$$\phi_k\cdot\phi_r = \phi_{kr} \text{ for } k, r \in \mathbf{Z}^+, \text{ i.e., } \{\phi_k | k \in \mathbf{Z}^+\}$$

is a mixing semigroup under composition.[2]

---

[2] This semigroup can be interpreted in terms of Tchebyschev polynomials (R. Adler and T. Rivlin [1]).

APPENDIX 15

SKEW-PRODUCTS

(See Definition 9.5)

Let (M, μ) and (M', μ') be two Lebesgue spaces. We denote by $(M \times M', \mu \times \mu')$ their direct product: the measure algebra $\hat{1}_{M \times M'}$ of $M \times M'$ is generated by the $A \times A'$, $A \in \hat{1}_M$, $A' \in \hat{1}_{M'}$ and we set $(\mu \times \mu')(A \times A') = \mu(A) \cdot \mu(A')$ for any $A \in \hat{1}_M$, $A' \in \hat{1}_{M'}$.

Assume that (M, μ, ϕ) is a dynamical system, and make correspond to every $m \in M$ an automorphism ψ_m: $M' \to M'$ such that $(m, m') \to \psi_m(m')$ is measurable for every $m \in M$, $m' \in M'$. *Then*

$$(\phi \times \{\psi\}): \; M \times M' \to M \times M',$$

defined by

$$(\phi \times \{\psi\})(m, m') = (\phi m, \psi_m m'),$$

is measurable and measure-preserving.

In fact, for every measurable set $F \in \hat{1}_{M \times M'}$ whose characteristic function is χ_F, we have:

$$\begin{aligned}(\mu \times \mu')((\phi \times \{\psi\})^{-1} F) &= \int_{M \times M'} \chi_F[(\phi \times \{\psi\})(m, m')] \mathrm{d}(\mu \times \mu') \\ &= \int_M \left[\int_{M'} \chi_F(\phi m, \psi_m m') \mathrm{d}\mu' \right] \mathrm{d}\mu .\end{aligned}$$

Since ψ_m is μ'-measure preserving, this may be written

$$(\mu \times \mu')((\phi \times \{\psi\})^{-1} F) = \int_M \left[\int_{M'} \chi_F(\phi m, m') \mathrm{d}\mu' \right] \mathrm{d}\mu .$$

Apply Fubini's theorem and observe that ϕ is measure-preserving:

$$(\mu\times\mu')((\phi\times\{\psi\})^{-1}F) = \int_{M'}\left[\int_M \chi_F(\phi m, m')d\mu\right]d\mu'$$

$$= \int_{M'}\left[\int_M \chi_F(m, m')\, d\mu\right]d\mu' = (\mu\times\mu')F \ .$$

The dynamical system $(M\times M', \mu\times\mu', \phi\times\{\psi\})$ *is called the skew-product of* (M, μ, ϕ) *and* (M', μ', ψ_m).

EXAMPLE A 15.1

If $\psi_m = \phi'$ is constant, that is, if $\phi\times\{\psi\}$ is defined by

$$(\phi\times\phi')(m, m') = (\phi m, \phi' m') \ ,$$

then the skew-product is the direct product of (M, μ, ϕ) and (M', μ', ϕ').

EXAMPLE A 15.2

Take $M = S^1 = \{x\,(\text{mod } 1)\}$ with the usual measure dx and ϕ: $S^1 \to S^1$ an ergodic translation:

$$\phi x = x + \omega \ (\text{mod } 1),$$

ω irrational. Now, *let* n *be an integer.* We take $M' = S^1 = \{y\ (\text{mod } 1)\}$ with the usual measure dy and we make correspond to each $x \in M = S^1$ a translation $\psi_{x,n}$: $M' \to M'$ defined by:

$$\psi_{x,n}(y) = y + nx \quad (\text{mod } 1) \ .$$

Thus, to any integer n corresponds the skew-product $(S^1\times S^1, dx\times dy, \phi\times\{\psi_{x,n}\})$.

Anzai [1] proved that the ergodic measure-preserving automorphisms $\phi\times\{\psi_{x,n}\}$ and $\phi\times\{\psi_{x,p}\}$, with ω irrational and $|n| \neq |p|$, are not isomorphic although they have the same spectral type and the same vanishing entropy.

APPENDIX 16

DISCRETE SPECTRUM OF CLASSICAL SYSTEMS

(See 9.13)

Let (M, μ, ϕ_t) be a classical flow and U_t the one-parameter group of unitary transformations induced by ϕ_t. The discrete component of the spectrum of U_t is called the *discrete spectrum.*

The dynamical systems constructed in the second part of the discrete spectrum theorem (see 9.13) are classical systems if the rank[1] of the abelian group of the proper values of U_t is finite.

All known examples of ergodic classical systems have discrete spectrum with finite rank and this rank is bounded from above by the dimension of the space.[2] Thus, the finiteness of this rank is a natural conjecture.

The proper functions of a classical system can be everywhere discontinuous (see an example due to Kolmogorov [1]), but if all of them are continuous, then the rank of the discrete spectrum is bounded from above by the first Betti number $b_1 = \dim H_1(M, \mathbf{R})$.

THEOREM A 16.1

If the proper functions of the induced unitary group U_t of an ergodic classical system (M, μ, ϕ_t) are continuous, then:

$$\textit{rank of the discrete spectrum} \leq b_1 .$$

This theorem is an obvious corollary of the following one:

---

[1] I. e., the maximum number of independent generators.

[2] This fact is general for systems with continuous proper functions (Avez [2]).

THEOREM A 16.2

Let (M, μ, ϕ_t) *be an ergodic classical system. Denote by C the subgroup of the discrete spectrum corresponding to continuous proper functions, then:*

$$\text{rank } C \leq b_1 .$$

Before proving this theorem we first introduce the winding numbers.

WINDING NUMBERS A 16.3

Let (M, μ, ϕ_t) be an ergodic classical system. The first homology group $H_1(M, \mathbf{R})$ has finite integral base: $\gamma_1, \ldots, \gamma_{b_1}$. Each γ_k is a closed curve which can be assumed differentiable. Let $a = \{\phi_t x | \ x \in M, \ 0 \leq t \leq T\}$ be an arc of an orbit of ϕ_t. We join the endpoints $\phi_T x$ and x with a geodesic arc β (geodesic in the sense of some Riemannian metric). Thus, $\gamma(T) = a\beta$ is a piecewise differentiable closed curve (see Figure A 16.3′) and there exist integers $n_k(T)$ such that:

$$\gamma(T) = n_1(T) \cdot \gamma_1 + \cdots + n_{b_1}(T) \cdot \gamma_{b_1} .$$

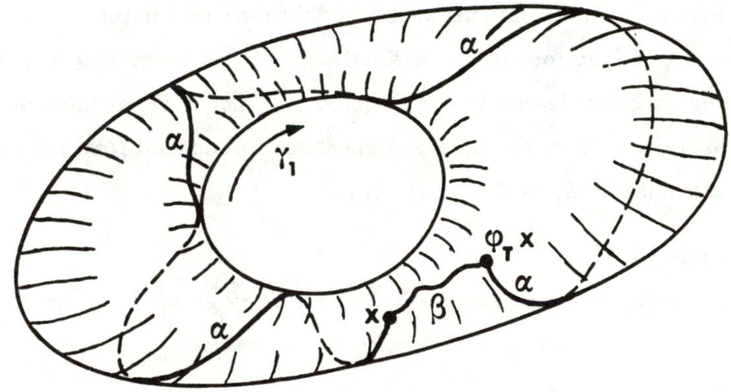

Figure A 16.3′

Let (ω_k) be the dual base of (γ_k) in the first cohomology group $H^1(M, \mathbf{Z})$, that is the closed one-forms satisfying:

$$\int_{\gamma_k} \omega_i = \begin{cases} 1 & \text{if } i = k \\ 0 & \text{if } i \neq k . \end{cases}$$

We obtain:

$$\int_{\gamma(T)} \omega_k = n_k(T) ,$$

which can be rewritten:

$$\int_0^T (\dot{\gamma}, \omega_k)\, dt + \int_\beta \omega_k = n_k(T) , \tag{A 16.4}$$

where $(\dot{\gamma}, \omega_k)$ is the value of the one-form ω_k for the infinitesimal generator $\dot{\gamma}$ of ϕ_t at point $\phi_t x$. Since length $\beta \leq$ diameter M, we have:

$$\lim_{T\to\infty} \frac{1}{T} \int_\beta \omega_k = 0 . \tag{A 16.5}$$

On the other hand, ergodicity implies (see 7.1):

$$\lim_{T\to\infty} \frac{1}{T} \int_0^T (\dot{\gamma}, \omega_k)\, dt = \int_M (\dot{\gamma}, \omega_k)\, d\mu , \tag{A 16.6}$$

for almost every initial point x, and the limit does not depend on x. Finally, from (A 16.4), (A16.5), and (A16.6) we deduce that:

$$\lim_{T\to\infty} \frac{n_k(T)}{T} = \int_M (\dot{\gamma}, \omega_k)\, d\mu \overset{\text{def}}{=} \mu_k \tag{A 16.7}$$

exists for almost every x and does not depend on x.

The numbers $\sigma_k = e^{2\pi i \mu_k}$, $k = 1, \ldots, b_1$, *generate a subgroup* $\mathcal{R}$ *of the circle group;* $\mathcal{R}$ *is called the group of the winding numbers.* It is readily seen that $\mathcal{R}$ does not depend on the base (γ_k). Thus, ϕ_t defines a real homology class:

$$\gamma = \mu_1 \cdot \gamma_1 + \cdots + \mu_{b_1} \cdot \gamma_{b_1} ;$$

the winding numbers μ_k define the "homological position" of a generic orbit. In other words, they define how an "average" orbit wanders around M.

This concept was first introduced by H. Poincaré [1] for flows on the torus T^2. Further investigations of systems:

$$\dot{x} = F(x, y) , \qquad \dot{y} = G(x, y)$$

on the torus T^2 are due to A. Denjoy [1] and C. L. Siegel.

Proof of Theorem A 16.2[3]

From its very construction, the rank of the winding numbers group is bounded from above by b_1, Therefore, Theorem (A 16.2) is a corollary of the following lemma.

LEMMA A 16.8

The subgroup C of the discrete spectrum corresponding to continuous proper functions is a subgroup of the winding numbers group.

Proof:

Let $f(x)$ be a nonvanishing continuous proper function of ϕ_t:

$$f(\phi_t x) = e^{2\pi i\lambda t} \cdot f(x) .$$

The function f is continuously differentiable with respect to the flow ϕ_t, that is:

$$(\dot{\gamma}(t), df(\phi_t x)) = 2\pi i\lambda \cdot e^{2\pi i\lambda t} \cdot f(x) ,$$

which may be written:

$$(\dot{\gamma}, df(x)) = 2\pi i\lambda \cdot f(x) \tag{A 16.9}$$

for $t = 0$. But ergodicity implies (see Theorem 9.12)

$$|f(x)| = \text{constant} \quad \text{a.e.},$$

that is to say

$$|f(x)| = \text{constant} \neq 0 \text{ everywhere},$$

because f is continuous. Thus, up to a constant, we have:

$$f(x) = e^{2\pi i\psi(x)} ,$$

where $\psi : M \to S^1$ is continuous and (A 16.9) reads:

$$(\dot{\gamma}, d\psi) = \lambda . \tag{A 16.10}$$

Therefore $d\psi$ defines a closed current of degree 1 in De Rham's sense.[4]

---

[3] See also I. M. Gelfand and Shapiro-Piatetski [1], S. Schwartzman [1].

[4] G. De Rham, *Variétés différentiables*, Hermann (Paris) 1955.

As it is well-known, such a current is homologous to a smooth closed one-form $[d\psi]$:

$$d\psi = [d\psi] + dh \; .$$

Assume that M is metrized with a Riemannian metric whose volume element is $d\mu$. Since ϕ_t is μ-measure-preserving, the infinitesimal generator $\dot{\gamma}$ of ϕ_t is co-closed ($\delta\dot{\gamma} = 0$). Then, relation (A 16.10) implies:

$$\int_M (\dot{\gamma}, [d\psi])\, d\mu = \int_M (\dot{\gamma}, d\psi)\, d\mu = \lambda \; .$$

Now it is sufficient, according to (A 16.3), to prove that $[d\psi]$ has integral periods. Let u: $[0, 1] \to M$ be an arbitrary smooth loop of M. Since $d\psi$ and $[d\psi]$ are homologous, we have:

$$\int_u [d\psi] = \int_u d\psi = \psi[u(1)] - \psi[u(0)] \in \mathbf{Z} \; .$$

(Q. E. D.)

COROLLARY A 16.11 (Arnold [2], [3]).

Let V be a compact orientable Riemannian manifold which is not a torus and the dimension of which is greater than one. If the geodesic flow on the unitary tangent bundle $M = T_1 V$ is ergodic, then the continuous proper functions reduce to constants.

Proof:

According to Lemma (A 16.8) it is sufficient to prove that every winding number vanishes. Under our topological assumptions, Gysin [1] proved that every closed one-form ω, which is not homologous to zero, is the lift of a closed one-form of V, say again ω. On the other hand, a winding number of the flow has the form (see A 16.7):

$$\mu = \int_{T_1 V} (\dot{\gamma}, \omega) \eta \wedge \sigma ,$$

where η (resp. σ) is the volume element of V (resp. the fiber S^{n-1}).

From

$$\int_{S^{n-1}} \dot{\gamma} \cdot \sigma = 0$$

we deduce:

$$\mu = \int_V \left(\int_{S^{n-1}} (\dot{\gamma}, \omega) \sigma \right) \eta = \int_V \left(\int_{S^{n-1}} \dot{\gamma} \sigma, \omega \right) \eta = 0 .$$

(Q. E. D.)

APPENDIX 17

SPECTRA OF κ-SYSTEMS

(See Theorem 11.5)

§A. Subalgebras of Measurable Sets

Let (M, μ) be a measure space. We denote by $\hat{1}$ the algebra of all the measurable sets and by $\hat{0}$ the algebra of the subsets of measure 0 or 1.

DEFINITION A 17.1

A subalgebra $\mathfrak{A}$ of measurable sets is a subset of $\hat{1}$ which is closed under the formation of complement and the denumerable union, and which contains M.

INCLUSION A 17.2

If $\mathfrak{A}_0$ and $\mathfrak{A}_1$ are subalgebras of $\hat{1}$, then $\mathfrak{A}_0 \subset \mathfrak{A}_1$ means that $\mathfrak{A}_0$ is a subalgebra of $\mathfrak{A}_1$, that is, that every element of $\mathfrak{A}_0$ is an element of $\mathfrak{A}_1$. The relation $\subset$ is a reflexive partial ordering on the family of subalgebras of $\hat{1}$.

INTERSECTION A 17.3

Let $(\mathfrak{A}_i)_{i \in I}$ be a family of subalgebras of $\hat{1}$. We denote by

$$\bigcap_{i \in I} \mathfrak{A}_i$$

the largest subalgebra of $\hat{1}$ which belongs to each $\mathfrak{A}_i$.

SUM A 17.4

Likewise, we denote by

$$\overline{\bigvee_{i \in I} \mathcal{A}_i}$$

the sum of the $\mathcal{A}_i$'s, that is, the smallest subalgebra of $\hat{1}$ which contains every $\mathcal{A}_i$.

THE SPACE $L_2(\mathcal{A})$. A 17.5

Let $\mathcal{A}$ be a subalgebra of $\hat{1}$. We denote by $L_2(\mathcal{A})$ the subspace of $L_2(M, \mu)$ generated by the characteristic functions of the elements of $\mathcal{A}$. The following properties are readily verified:

$$\mathcal{A} \subset \mathcal{B} \text{ implies } L_2(\mathcal{A}) \subset L_2(\mathcal{B}) \;,$$

$$L_2\Big(\bigcap_{i \in I} \mathcal{A}_i\Big) = \bigcap_{i \in I} L_2(\mathcal{A}_i) \;,$$

$$L_2\Big(\overline{\bigvee_{i \in I} \mathcal{A}_i}\Big) = \overline{\bigcup_{i \in I} L_2(\mathcal{A}_i)} \;,$$

$L_2(\hat{0}) = H_0$, one-dimensional space of the constants.

§ B. Spectra of K-Systems

We next prove the following theorem (see Theorem 11.5).

THEOREM A 17.6

A K-system (M, μ, ϕ) has denumerably multiple Lebesgue spectrum.

Recall that there exists (see Definition 11.1) a subalgebra $\mathcal{A}$ of $\hat{1}$ such that:

(A 17.7)

$$\hat{0} = \bigcap_{n=-\infty}^{\infty} \phi^n\mathcal{A} \cdots \subset \phi^{-1}\mathcal{A} \subset \mathcal{A} \subset \phi\mathcal{A} \subset \cdots \subset \overline{\bigvee_{n=-\infty}^{\infty} \cdots \phi^n\mathcal{A}} = \hat{1} \,.$$

The proof breaks up into several lemmas.

LEMMA A 17.8

Let U be the unitary operator induced by ϕ. If $H = L_2(\mathcal{A})$ then:

$$H_0 = \bigcap_{n=-\infty}^{\infty} U^n H \subset \cdots \subset UH \subset H \subset \cdots \subset \overline{\bigcup_{n=-\infty}^{\infty} U^n H} = L_2(M, \mu) .$$

Let us denote by $H \ominus H_0 = H'$ the orthocomplement of H_0 in H; this may be written again:

$$\{0\} = \bigcap_{n=-\infty}^{\infty} U^n H' \subset \cdots \subset UH' \subset H' \subset U^{-1}H' \subset \cdots$$

$$\subset \overline{\bigcup_{n=-\infty}^{\infty} U^n H'} = L_2' = L_2(M, \mu) \ominus H_0 .$$

Proof:

Let A be an element of $\mathcal{C}$, and $\mathcal{X}_A$ its characteristic function. Then:

$$U\mathcal{X}_A(x) = \mathcal{X}_A(\phi x) = \begin{cases} & \text{if } \phi(x) \notin A \\ & \text{if } \phi(x) \in A , \end{cases}$$

that is,

$$U\mathcal{X}_A(x) = \begin{cases} 0 & \text{if } x \notin \phi^{-1}A \\ 1 & \text{if } x \in \phi^{-1}A . \end{cases}$$

Thus, $U\mathcal{X}_A = \mathcal{X}_{\phi^{-1}A}$ and, according to the definition of $L_2(\mathcal{C})$, we have:

$$UL_2(\mathcal{C}) = L_2(\phi^{-1}\mathcal{C}) .$$

Now, the lemma is a direct corollary of properties (A 17.5).

LEMMA A 17.9

U has Lebesgue spectrum, the multiplicity of which is equal to

$$\dim(H \ominus UH) .$$

Proof:

Let $\{h_i\}$ be a complete orthonormal basis of $H' \ominus UH'$. Denote by $\mathcal{H}_i$ the closure of the subspace spanned by $h_i, Uh_i, \ldots$. From their very construction, the $U^j h_i$'s, and so the $\mathcal{H}_i$'s are orthogonal to each other. From Lemma (A 17.8) we have $\bigcap_{n=-\infty}^{\infty} U^n H' = \{0\}$, therefore $\{U^j h_i\}$ is a complete system of H':

(A 17.10)
$$H' = \oplus \sum_i \mathcal{H}_i \, .$$

On the other hand, the relation:

$$\overline{\bigcup_{n=-\infty}^{\infty} U^n H'} = L_2'$$

can be rewritten:

$$\overline{\bigcup_{n=0}^{\infty} U^{-n} H'} = L_2' \, .$$

This relation and (A 17.10) yield:

$$L_2' = \overline{\bigcup_{n=0}^{\infty} U^{-n}\Big(\oplus \sum_i \mathcal{H}_i\Big)} = \oplus \sum_i \overline{\bigcup_{n=0}^{\infty} U^{-n} \mathcal{H}_i} \, .$$

Let us set:

(A 17.11)
$$H_i = \overline{\bigcup_{n=0}^{\infty} U^{-n} \mathcal{H}_i} \, ,$$

this can be rewritten

(A 17.12)
$$L_2' = \oplus \sum_i H_i \, .$$

According to (A 17.11), from the basis $\{h_i, Uh_i, \ldots\}$ of $\mathcal{H}_i$ we obtain a complete orthonormal basis of H_i, namely:

$$\{e_{i,j} \overset{\text{def}}{=} U^j h_i \,|\, j \in \mathbf{Z}\} \, .$$

Furthermore:

$$Ue_{i,j} = U(U^j h_i) = U^{j+1} h_i = e_{i,j+1}$$

for every i and j.

Together with (A 17.12), this proves that U has Lebesgue spectrum, the multiplicity of which is equal to the cardinality of $\{H_i\}$, *that is*

$$\dim(H' \ominus UH') = \dim(H \ominus UH) \, .$$

LEMMA A 17.13

$$\dim(H \ominus UH) = \infty .$$

Proof:

Lemma (A 17.8) implies:

$$\cdots \leq \dim UH \leq \dim H \leq \dim U^{-1}H \leq \cdots .$$

This proves $\dim H = \infty$ and $UH \neq H$, since $\dim H < \infty$ would imply $\dim U^n H = \dim U^{n+1} H = \dim\{0\}$ for n large enough, that is $H = 0$. Similarly, $\dim UH = \infty$. Since $UH \neq H$, there exists some non-vanishing function $f \in H \ominus UH$. Let us denote by F the support of f:

$$F = \{m \mid m \in M, f(m) \neq 0\} .$$

We have $F \in \mathfrak{A}$ and $\mu(F) > 0$ for $f \in L_2(\mathfrak{A})$, and the space

$$L = \{g \cdot \mathfrak{X}_F \mid g \in H\} \quad (\mathfrak{X}_F\text{: characteristic function of } F)$$

has infinite dimension for $\dim H = \infty$. Similarly, as $\mu(F) > 0$ and $\dim UH = \infty$, the space $L_1 = \{g \cdot \mathfrak{X}_F \mid g \in UH\}$ has infinite dimension. Let us set $L_0 = L \ominus L_1$ and take $g \cdot \mathfrak{X}_F \in L_0$ and $h \in UH$, then we have:

$$< g\mathfrak{X}_F \mid h > = < g\mathfrak{X}_F \mid h\mathfrak{X}_F > = 0 .$$

Thus $L_0 \subset H \ominus UH$, and it is sufficient to prove that $\dim L_0 = \infty$.

As L_1 is infinite-dimensional and $\mu(F) > 0$, there exists a sequence of bounded real-valued functions: $h_1, h_2, \ldots \in UH$ such that the $\mathfrak{X}_F \cdot h_1$, $\mathfrak{X}_F h_2, \ldots \in L_1$ are linearly independent. Since f does not vanish in F, we find that the $fh_1, fh_2, \ldots \in L$ are linearly independent. But they belong to L_0 because $h \in UH$ implies:

$$< fh_k \mid h\mathfrak{X}_F > = < f \mid h_k h > = 0 \text{ for every } k$$

($h_k \cdot h \in UH$ is orthogonal to f). Thus, there exist infinitely many linearly independent functions in L_0.

(Q. E. D.)

The preceding lemmas prove Theorem (A 17.6).

APPENDIX 18

CONDITIONAL ENTROPY OF A PARTITION α WITH RESPECT TO A PARTITION β

(See Section 12, Chapter 2)

§ A. Measurable Partitions

DEFINITION A 18.1

Let (M, μ) be a measure space. *A partition* $\alpha = \{A_i\}_{i \in I}$ *of* M *is a collection of nonempty, nonintersecting measurable sets that cover* M:

$$\mu(A_i \cap A_j) = 0 \text{ if } i \neq j, \quad \mu(M - \bigcup_i A_i) = 0 \ .$$

A partition α *is said to be measurable if there exists a countable system* $\{B_j\}_{j \in J}$ *of measurable sets such that:*

(1) *each* B_j *is a sum of elements of* α ;

(2) *for any two elements* A_i, A_j, *of* α *there exists a* B_k *such that either* $A_i \subset B_k$, $A_j \not\subset B_k$, *or* $A_i \not\subset B_k$, $A_j \subset B_k$.

A finite or countable partition is clearly measurable.

DEFINITION A 18.2

From the very definition, it follows that we can remove the elements of measure zero from a partition. More generally, *two partitions* α *and* β *will be identified:* $\alpha = \beta$ (mod 0) *if their elements coincide up to some sets of measure zero.*

In the future we delete (mod 0).

DEFINITION A 18.3

A partition β *is said to be a refinement of a partition* α: $\alpha \leqq \beta$ *if* *every element B of β is a subset of some element A of α:* $\mu(B - B \cap A) = 0$.

DEFINITION A 18.4

Let $\{\alpha_i\}_{i \in I}$ be a family of measurable partitions. We define their sum:

$$\alpha = \bigvee_{i \in I} \alpha_i$$

as the smallest partition which contains every α_i. In other words:

$$\alpha = \left\{ \bigcap_{j \in I} A_j \mid A_j \in \alpha_j \text{ for all } j \right\}.$$

The operation $\bigvee$ is commutative and associative and

$$\alpha \leq \alpha', \quad \beta \leq \beta' \text{ imply } \alpha \vee \beta \leq \alpha' \vee \beta'.$$

DEFINITION A 18.5

Given an arbitrary measurable partition α, we denote by $\mathfrak{M}(\alpha)$ the subalgebra of the algebra $\hat{1}$ that consists of the sets that are sums of elements of α. The algebra $\mathfrak{M}(\alpha)$ is called the algebra generated by α.

It turns out that for every subalgebra $\mathfrak{A}$ of $\hat{1}$, there exists a measurable partition α such that:

$$\mathfrak{A} = \mathfrak{M}(\alpha) .$$

One verifies at once:

$$\alpha = \beta \Longleftrightarrow \mathfrak{M}(\alpha) = \mathfrak{M}(\beta)$$

$$\alpha \leq \beta \Longleftrightarrow \mathfrak{M}(\alpha) \subset \mathfrak{M}(\beta)$$

$$\mathfrak{M}\left(\bigvee_{i \in I} \alpha_i \right) = \bigvee_{i \in I} \mathfrak{M}(\alpha_i) .$$

(See A. N. Kolmogorov [3] and V. Rohlin [3]).

§ B. Entropy of α Given β

Let $\alpha = \{A_i \mid i = 1, \dots, r\}$ and $\beta = \{B_j \mid j = 1, \dots, s\}$ be two finite measurable partitions. We can assume, without losing generality, that any element A_i or B_j has positive measure (see A 18.2).

DEFINITION A 18.6

Let $z(t)$ be the function over $[0, 1]$ defined by:

$$z(t) = \begin{cases} -t \log t & \text{if } 0 < t \leq 1 \\ 0 & \text{if } t = 0. \end{cases}$$

The conditional entropy of α with respect to β is:

$$h(\alpha/\beta) = \sum_j \mu(B_j) \sum_i z(\mu(A_i/B_j)),$$

where

$$\mu(A_i/B_j) = \frac{\mu(A_i \cap B_j)}{\mu(B_j)}$$

is the conditional measure of A_i relative to B_j.

We turn next to the proof of Theorem (12.5) which we reformulate.

THEOREM 12.5

Let $\alpha = \{A_i\}$, $\beta = \{B_j\}$, $\gamma = \{C_k\}$ be finite measurable partitions. Then:

(12.6) $\quad h(\alpha/\beta) \geq 0$ with equality if, and only if $\alpha \leq \beta$;

(12.7) $$h(\alpha \vee \beta/\gamma) = h(\alpha/\gamma) + h(\beta/\alpha \vee \gamma);$$

(12.8) $$\alpha \leq \beta \Longrightarrow h(\alpha/\gamma) \leq h(\beta/\gamma);$$

(12.9) $$\beta \leq \gamma \Longrightarrow h(\alpha/\gamma) \geq h(\alpha/\beta);$$

(12.10) $$h(\alpha \vee \beta/\gamma) \leq h(\alpha/\gamma) + h(\beta/\gamma).$$

Proof:

Proof of (12.6) is left to the reader as an easy exercise.

The elements of $\alpha \vee \beta$ and $\alpha \vee \gamma$ are, respectively, of the form: $A_i \cap B_j$ and $A_i \cap C_k$. Therefore

$$h(\alpha \vee \beta/\gamma) = -\sum_{i,j,k} \mu(A_i \cap B_j \cap C_k) \operatorname{Log} \mu(A_i \cap B_j/C_k).$$

But we have

$$\mu(A_i \cap B_j / C_k) = \frac{\mu(A_i \cap B_j \cap C_k)}{\mu(C_k)} = \frac{\mu(A_i \cap C_k)}{\mu(C_k)} \frac{\mu(A_i \cap B_j \cap C_k)}{\mu(A_i \cap C_k)}$$

$$= \mu(A_i/C_k) \cdot \mu(B_j/A_i \cap C_k)$$

and we deduce relation (12.7):

$$h(\alpha \vee \beta/\gamma) = - \sum_{i,j,k} \mu(A_i \cap B_j \cap C_k) \text{ Log } \mu(A_i/C_k)$$

$$- \sum_{i,j,k} \mu(A_i \cap B_j \cap C_k) \text{ Log } \mu(B_j/A_i \cap C_k)$$

$$= - \sum_{i,k} \mu(A_i \cap C_k) \text{ Log } \mu(A_i/C_k) -$$

$$- \sum_{i,j,k} \mu(B_j \cap (A_i \cap C_k)) \text{ Log } \mu(B_j/A_i \cap C_k)$$

$$= h(\alpha/\gamma) + h(\beta/\alpha \vee \gamma) .$$

Let us prove relation (12.8): If $\alpha \leq \beta$, then $\alpha \vee \beta = \beta$ and relations (12.6) and (12.7) imply:

$$h(\beta/\gamma) = h(\alpha/\gamma) + h(\beta/\alpha \vee \gamma) \geq h(\alpha/\gamma) .$$

Let us prove relation (12.9): Since $\Sigma_k \mu(C_k/B_j) = 1$ and $\mu(C_k/B_j) \geq 0$, the concavity of $z(t)$ implies:

$$\sum_k z(\mu(A_i/C_k)) \cdot \mu(C_k/B_j) \leq z \left[\sum_k \mu(A_i/C_k) \cdot \mu(C_k/B_j) \right] .$$

Since $\beta \leq \gamma$, each B_j is the disjoint union of some C_k's; therefore we have:

$$\sum_k \mu(A_i/C_k) \cdot \mu(C_k/B_j) = \sum_{k'} \frac{\mu(A_i \cap C_{k'})}{\mu(C_{k'})} \frac{\mu(C_{k'})}{\mu(B_j)} = \mu(A_i/B_j) ,$$

where the sum extends over those $C_{k'}$'s belonging to B_j. We deduce:

$$\sum_k z(\mu(A_i/C_k)) \cdot \mu(C_k/B_j) \leq z[\mu(A_i/B_j)] .$$

Multiplying both members by $\mu(B_j)$ and summing over i and j yields (12.9).

Finally, (12.10) is a consequence of (12.7) and (12.9): $\alpha \vee \gamma \geq \gamma$ implies

$$h(\beta/\alpha \vee \gamma) \leq h(\beta/\gamma)$$

and

$$h(\alpha \vee \beta/\gamma) = h(\alpha/\gamma) + h(\beta/\alpha \vee \gamma) \leq h(\alpha/\gamma) + h(\beta/\gamma) .$$

The preceding definitions and properties extend to denumerable measurable partitions (see Rohlin and Sinai [5]).

APPENDIX 19

ENTROPY OF AN AUTOMORPHISM

(See Theorem 12.26)

The purpose of this appendix is to prove the following theorem due to Kolmogorov.[1]

THEOREM A 19.1

If ϕ possesses a generator α, then $h(\phi) = h(\alpha, \phi)$:

The proof breaks into several lemmas. Denote by F the set of all finite measurable partitions. Given $\alpha, \beta \in F$ we write $|\alpha, \beta| = h(\alpha \mid \beta) + h(\beta \mid \alpha)$.

LEMMA 19.2

$|\alpha, \beta|$ *is a distance on* F.

Proof:

It is clear that $|\alpha, \beta| \geq 0$. From formula (12.6) of Chapter 2 we deduce:

$|\alpha, \beta| = 0 \Longrightarrow h(\alpha \mid \beta) = h(\beta \mid \alpha) = 0 \Longrightarrow \alpha \leq \beta$ and $\beta \leq \alpha \Longrightarrow \alpha = \beta$.

It is also evident that $|\alpha, \beta| = |\beta, \alpha|$. According to (12.11), (12.12), and (12.9) we have:

$$h(\alpha/\gamma) = h(\alpha \vee \gamma) - h(\gamma) \leq h(\alpha \vee \beta \vee \gamma) - h(\beta \vee \gamma) + h(\beta \vee \gamma) - h(\gamma)$$
$$= h(\alpha/\beta \vee \gamma) + h(\beta/\gamma) \leq h(\alpha/\beta) + h(\beta/\gamma)$$

[1] The proof follows Rohlin [4].

and symmetrically:

$$h(\gamma/\alpha) \leq h(\beta/\alpha) + h(\gamma/\beta) \, .$$

Addition yields:

$$|\alpha, \gamma| \leq |\alpha, \beta| + |\beta, \gamma| \, .$$

LEMMA A 19.3

Given ϕ, $h(\alpha, \phi)$ is a continuous function on F in its argument α. More precisely:

$$|h(\alpha, \phi) - h(\beta, \phi)| \leq |\alpha, \beta| \, .$$

Proof:

Given $\alpha, \beta \in F$, we set:

$$\alpha_n = \alpha \vee \phi\alpha \cdots \vee \phi^{n-1}\alpha; \quad \beta_n = \beta \vee \cdots \vee \phi^{n-1}\beta \, .$$

From (12.11) of Chapter 2 follows:

$$h(\beta_n/\alpha_n) - h(\alpha_n/\beta_n) = [h(\alpha_n \vee \beta_n) - h(\alpha_n)] - [h(\alpha_n \vee \beta_n) - h(\beta_n)]$$

$$= h(\beta_n) - h(\alpha_n) \, .$$

Since $h(\ |\ ) \geq 0$, we deduce:

$$|h(\beta_n) - h(\alpha_n)| \leq h(\beta_n/\alpha_n) + h(\alpha_n/\beta_n) \, .$$

On the other hand, from (12.10) of Chapter 2 follows:

$$h(\alpha_n/\beta_n) = h(\alpha \vee \cdots \vee \phi^{n-1}\alpha/\beta_n) \leq h(\alpha/\beta_n) + \cdots + h(\phi^{n-1}\alpha/\beta_n) \, .$$

Similarly, from (12.9) and because $\beta, \ldots, \phi^{n-1}\beta \leq \beta_n$, we have:

$$h(\alpha_n/\beta_n) \leq h(\alpha/\beta) + h(\phi\alpha/\phi\beta) + \cdots + h(\phi^{n-1}\alpha/\phi^{n-1}\beta) = nh(\alpha/\beta) \, .$$

Symmetrically:

$$h(\beta_n/\alpha_n) \leq n \cdot h(\alpha/\beta) \, .$$

Addition yields:

$$|h(\beta_n) - h(\alpha_n)| \leq n[h(\alpha/\beta) + h(\beta/\alpha)] = n \cdot |\alpha, \beta| \, .$$

Dividing both sides of this inequality by n and passing to the limit as $n \to \infty$, we obtain Lemma (A 19.3).

LEMMA A 19.4

If $\alpha_1, \alpha_2, \ldots$ is a sequence of finite partitions such that

$$\alpha_1 \leq \alpha_2 \leq \cdots \leq \alpha_n \leq \alpha_{n+1} \leq \cdots ,$$

$$\overline{\bigvee_{n=1}^{\infty} \mathfrak{M}(\alpha_n)} = \hat{1} .$$

then the set B of partitions $\beta \in F$, such that $\beta \leq \alpha_n$ for at least one value of n, is everywhere dense in F.

Proof:

We need to prove that for every finite partition α and every $\delta > 0$ there exist an n and a $\beta \in B$ such that:

$$\beta \leq \alpha_n, \quad |\alpha, \beta| < \delta .$$

Let $A_1, \ldots, A_m$ be the elements of α. Since

$$\bigvee_{n=1}^{\infty} \mathfrak{M}(\alpha_n)$$

is dense in $\hat{1}$, for every $\delta' > 0$ there exist an n and subsets $A_1', \ldots, A_{m-1}' \in \mathfrak{M}(\alpha_n)$ such that:

$$\mu(A_i \cup A_i' - A_i \cap A_i') < \delta', \qquad i = 1, \ldots, m-1 .$$

Let us denote by β the partition of M into sets $B_1, \ldots, B_m$ defined by:

$$B_1 = A_1', \; B_i = A_i' - A_i' \cap (A_1' \cup \cdots \cup A_{i-1}'), \qquad i = 2, \ldots, m-1,$$

$$B_m = M - (A_1' \cup \cdots \cup A_{m-1}') .$$

It is clear that $\beta \leq \alpha_n$. On the other hand:

$$|\alpha, \beta| = h(\alpha/\beta) + h(\beta/\alpha) =$$

$$- \sum_i \mu(A_i) \sum_k \frac{\mu(A_i \cap B_k)}{\mu(A_i)} \operatorname{Log}\left[\frac{\mu(A_i \cap B_k)}{\mu(A_i)}\right]$$

$$- \sum_k \mu(B_k) \sum_i \frac{\mu(B_k \cap A_i)}{\mu(B_k)} \operatorname{Log}\left[\frac{\mu(B_k \cap A_i)}{\mu(B_k)}\right]$$

$$= -2 \sum_{i,k} \mu(A_i \cap B_k) \operatorname{Log} \mu(A_i \cap B_k) + \sum_i \mu(A_i) \operatorname{Log} \mu(A_i)$$

$$+ \sum_k \mu(B_k) \operatorname{Log} \mu(B_k) \ .$$

These formulas show that $|\alpha, \beta|$ depends continuously on $A_1', \ldots, A_{m-1}'$ and vanishes when $A_1' = A_1, \ldots, A_{m-1}' = A_{m-1}$. Therefore, if δ' is small enough, then $|\alpha, \beta| < \delta$.

Proof of Kolmogorov theorem:

Assume that ϕ possesses a generator α. We set, for $\lambda \in F$ and $q = 0, 1, \ldots$:

$$\tilde{\lambda}_q = \phi^{-q+1} \lambda \vee \cdots \vee \lambda \vee \phi\lambda \vee \cdots \vee \phi^{q-1} \lambda \ .$$

We have:

$$\tilde{\alpha}_1 \leq \tilde{\alpha}_2 \leq \cdots \leq \tilde{\alpha}_n \leq \tilde{\alpha}_{n+1} \leq \cdots$$

$$\overline{\bigvee_{n=1}^{\infty} \mathfrak{M}(\tilde{\alpha}_n)} = \hat{1} .$$

From Lemma (A 19.4) it follows that the set B' of partitions $\beta \in F$ such that $\beta \leq \tilde{\alpha}_n$ for at least one value of n is everywhere dense in F. Let β be an element of B'. Clearly:

$$\tilde{\beta}_m \leq \widetilde{(\tilde{\alpha}_n)}_m = \tilde{\alpha}_{n+m-1} \ .$$

Therefore, from (12.12) of Chapter 2 follows:

$$h(\tilde{\beta}_m) \leq h(\tilde{\alpha}_{n+m-1})$$

and

$$\frac{h(\tilde{\beta}_m)}{m} \leq \frac{h(\tilde{a}_{n+m-1})}{n+m-1} \cdot \frac{n+m-1}{m}$$

Now, observe that:

$$\frac{h(\tilde{\lambda}_q)}{q} = \frac{h(\phi^{-q+1}\lambda \vee \cdots \vee \lambda \vee \cdots \vee \phi^{q-1}\lambda}{q}$$

$$= \frac{h(\lambda \vee \cdots \vee \phi^{2q-2}\lambda)}{2q-1} \cdot \frac{2q-1}{q} \to 2h(\lambda, \phi), \quad \text{as } q \to \infty .$$

Thus, passing to the limit as $m \to +\infty$, we obtain

$$h(\beta, \phi) \leq h(a, \phi) .$$

Recall that B' is everywhere dense in F and that $h(\beta, \phi)$ is continuous in β (Lemma A 19.3), then:

$$h(a, \phi) \geq \sup_{B'} h(\beta, \phi) = \sup_{F} h(\beta, \phi) = h(\phi) ,$$

that is:

$$h(a, \phi) = h(\phi) .$$

APPENDIX 20

EXAMPLES OF RIEMANNIAN MANIFOLDS WITH NEGATIVE CURVATURE

(See 14.1, Chapter 3)

Consider the proper affine group G of the real line $\{t \mid t \in \mathrm{R}\}$. An element g of G has the form:

$$g: \; t \to yt + x, \quad x, y \in \mathrm{R}, \quad y > 0,$$

and can be denoted by (x, y).

Given $g' = (x', y')$, we obtain:

$$g'(g(t)) = y'(yt + x) + x' = y'yt + y'x + x' .$$

Therefore, if we denote the group operation by $\perp$ this may be written:

$$(x', y') \perp (x, y) = (y'x + x', y'y) .$$

The neutral element is $e = (0, 1)$ and the inverse of (x, y) is $(-xy^{-1}, y^{-1})$. Both $\perp$ and $g \to g^{-1}$ are smooth operations. Thus, G is a Lie group that is diffeomorphic to the upper half-plane $\{(x, y) \mid y > 0\}$. Now we turn G into a Riemannian manifold.

THEOREM A 20.1. THE RIEMANNIAN METRIC OF G

The left-invariant metric of G which reduces to

$$ds^2 = dx^2 + dy^2$$

at the neutral element $e = (0, 1)$ is:

$$ds^2 = \frac{dx^2 + dy^2}{y^2} .$$

Proof:

To any element $X = (x, y)$ of G corresponds the left translation L_X:

$$L_X(U) = x \perp U, \quad \text{where } U = (u, v) \in G .$$

We have:

$$L_{x^{-1}}(U) = \left(\frac{u-x}{y}, \frac{v}{y}\right),$$

the tangent mapping of which is:

(A 20.2) $$L^*_{x^{-1}}\xi = \begin{pmatrix} y^{-1}\xi_1 \\ y^{-1}\xi_2 \end{pmatrix}, \quad \text{if } \xi = \begin{pmatrix} \xi_1 \\ \xi_2 \end{pmatrix} \in TG_X .$$

Define a metric over the Lie algebra TG_e by setting:

$$<\phi \mid \phi>_e = (\phi_1)^2 + (\phi_2)^2, \quad \phi = \begin{pmatrix} \phi_1 \\ \phi_2 \end{pmatrix} \in TG_e .$$

This defines a left-invariant metric at each point X:

$$<\xi \mid \xi>_X = <L^*_{x^{-1}}\xi \mid L^*_{x^{-1}}\xi>_e .$$

Therefore, if $X = (x, y)$, (A 20.2) implies:

$$<\xi \mid \xi>_X = \frac{(\xi_1)^2 + (\xi_2)^2}{y^2} .$$

In other words, the metric is:

(A 20.3) $$ds^2 = \frac{dx^2 + dy^2}{y^2} .$$

DEFINITION A 20.4

The upper half-plane G *endowed with the metric (A 20.3) is called the Lobatchewsky-Poincaré plane.*

It can be useful to represent a point (x, y) of G by the complex number $z = x + iy$.

THEOREM A 20.5. THE ISOMETRIES OF G

The symmetry $(x, y) \to (-x, y)$ and the homographies:

(A20.6) $$z \to z' = \frac{az+b}{cz+d}\ ; \quad a, b, c, d \in \mathbf{R},\ ad-bc = 1$$

preserve the metric (A 20.3).

Proof:

Proof is purely computational and easy if one observes that:

$$ds^2 = \frac{-4\,dz\,d\bar{z}}{(z-\bar{z})^2}, \quad \text{where } \bar{z} = x-iy\,.$$

THEOREM A 20.7. ANGLES

The angles of metric (A20.3) coincide with Euclidean angles.

Consequently, words such as "orthogonal," and so on can be used unambiguously.

Proof:

$$\frac{dx^2+dy^2}{y^2} \quad \text{is proportional to } dx^2+dy^2\,.$$

THEOREM A 20.8. GEODESICS

The geodesics of (A 20.3) are the straight lines: x = constant, $y > 0$ *and the upper half-circles centered on* ox. *In particular, there exists one, and only one, geodesic passing through two given distinct points.*

Proof:

Let ab be a segment of $x = 0,\ y > 0$. For any arc γ joining a and b we have:

$$\int_\gamma ds^2 = \int_a^b \frac{dx^2+dy^2}{y^2} \geq \int_a^b \frac{dy^2}{y^2} = \int_{ab} ds^2\,.$$

This proves that $x = 0,\ y > 0$ is a geodesic.

An image of this geodesic under any isometry (A 20.6) is still a geodesic. We obtain so all the upper half-circles centered on ox and the half-straight lines x = constant, $y > 0$. In fact we obtained all the geodesics for, given a vector $u \in T_1 G$, there exists a half-circle centered on ox (or a parallel to oy) which is tangent to $\mathbf{u}$.

THEOREM A 20.9. CURVATURE

The Gaussian curvature of (A 20.3) is equal to -1.

Proof:

The Gaussian curvature K is constant, for the metric is invariant under a transitive group of isometries. The Gauss-Bonnet formula applied to a geodesic triangle $\Delta = ABC$ gives:

$$\hat{A} + \hat{B} + \hat{C} = \pi + \iint_{\Delta} K \cdot d\sigma = \pi + K \cdot \text{area}\ \Delta .$$

The particular case of Figure (A 20.10) gives $\hat{A} = \hat{B} = \hat{C} = 0$. As the element of area is $d\sigma = (dx\,dy)/y^2$, we obtain:

$$\text{area}\ \Delta = 2 \int_0^r dx \int_{r \sin\theta}^{\infty} \frac{dy}{y^2} = \pi .$$

We conclude that $K = -1$.

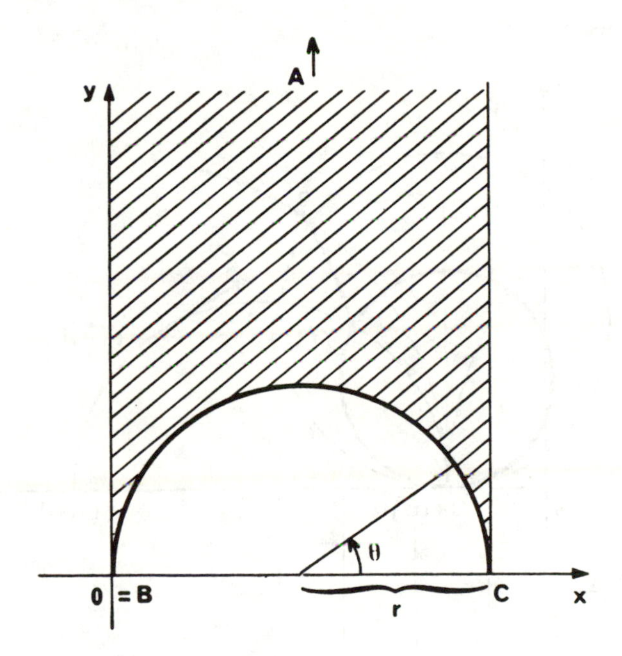

Figure A 20.10

THEOREM A 20.11. ASYMPTOTIC GEODESICS

Let $\gamma(\mathbf{u}, t) = \gamma(t)$ be a geodesic parametrized by arc length t, and $g \in G$. The geodesic passing through g and $\gamma(t_1)$ has a limit position as $t_1 \to +\infty$ (resp. $-\infty$). This limit position is the geodesic passing through g and the intersection $\gamma(+\infty)$ (resp. $\gamma(-\infty)$) of γ with ox.

Geodesics emanating from $\gamma(+\infty)$ (resp. $\gamma(-\infty)$) are called the positive (resp. negative) asymptotes to γ.

Proof:

Let $\gamma(t_1)$ be a point of γ. The geodesic passing through g and $\gamma(t_1)$ is a circle centered on ox, possibly reduced to a straight line (A 20.8).

From the very definition of the metric (A 20.3), $\gamma(t_1)$ runs to ox as $t_1 \to +\infty$ (resp. $-\infty$), that is $\gamma(t_1)$ converges to the intersection $\gamma(+\infty)$ (resp. $\gamma(-\infty)$) of γ with ox (see Figure A 20.12). Thus our geodesic has a limit position, namely the upper half-circle centered on ox and passing through g and $\gamma(+\infty)$ (resp. $\gamma(-\infty)$). Consequently, this limit position is a geodesic.

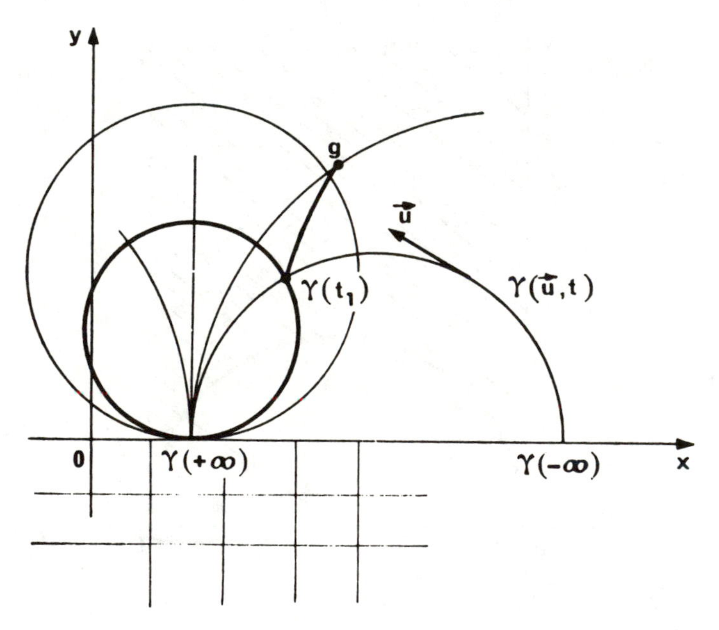

Figure A 20.12

DEFINITION A 20.13. HOROCYCLES[1]

The orthogonal trajectories of the positive (resp. negative) asymptotes to γ are called the positive (resp. negative) horocycles of γ.

THEOREM A 20.14.

The positive (resp. negative) horocycles of γ are the Euclidean circles of G which are tangent to $y = 0$ at $\gamma(+\infty)$ (resp. $\gamma(-\infty)$). In particular, the straight lines $y = C > 0$ are horocycles. They are positive horocycles of the axis oy $(y \to \infty)$.

Proof:

The positive (resp. negative) asymptotes to γ form the upper part of the pencil of circles that are orthogonal to $y = 0$ at $\gamma(+\infty)$ (resp. $\gamma(-\infty)$). Theorem (A 20.14) follows at once from the elementary properties of conjugate pencils of circles.

The points $\gamma(+\infty)$ and $\gamma(-\infty)$, which do not belong to G, have to be removed.

THEOREM A 20.15. RIEMANNIAN CIRCLES

The Riemannian circles of (A 20.1) centered at m form the upper part of the pencil of circles whose radical axis is ox and whose Poncelet points consist in m and the symmetric m′ of m with respect to ox.

Proof:

The Riemannian circles centered at m are the orthogonal trajectories of the geodesics emanating from m. This family of geodesics is nothing but the upper part of the pencil of circles passing through m and m'.

(Q. E. D.)

In particular, the power of any point d of ox with respect to one of these Riemannian circles centered at m is:

$$(dm)^2 = (dI)^2 + (Im)^2$$

(see Figure A 20.17).

[1] Notion due to Lobatchewsky (in Greek: "horos" = horizon).

THEOREM A 20.16

Horocycles are Riemannian circles the radii of which are infinite and the centers of which are at infinity (on $y = 0$*).*

Proof:

Consider the Riemannian circle passing through a fixed point n of a geodesic γ and centered at $m \in \gamma$ (see Figure A 20.17). If m moves to infinity along γ, that is, if m converges to ox, then $mm' \to 0$. Therefore, the power of any point of ox with respect to our circle tends to zero. Thus, our circle has a limit position which is the circle tangent to ox at $\gamma(+\infty)$ and which passes through n. Theorem (A 20.16) shows that this limit position is an horocycle. Conversely, any horocycle is obtained from the above construction. (Q. E. D.)

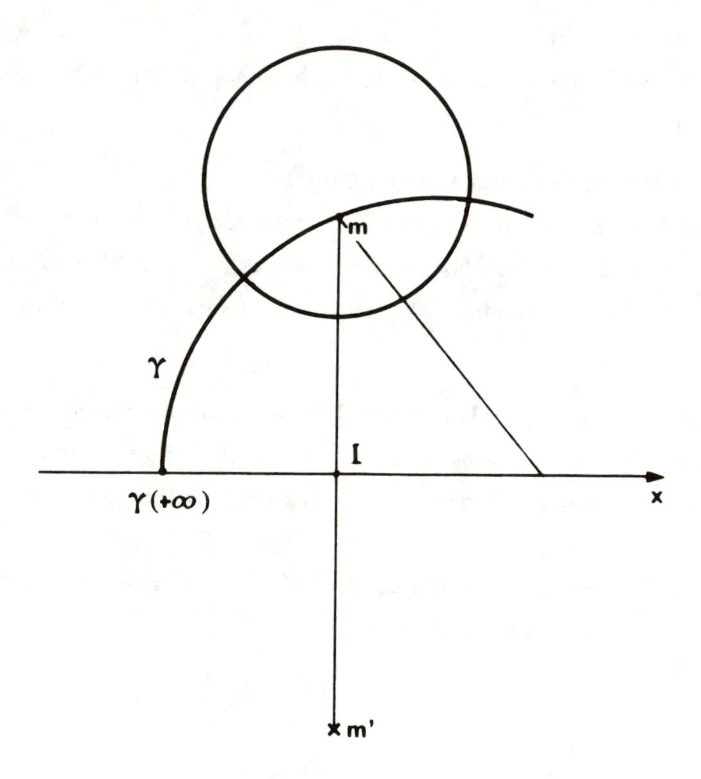

Figure A 20.17

THEOREM A 20.18

Let $\gamma(\mathrm{u}, t)$ and $\gamma'(\mathrm{u}', t)$ be two geodesics which are positively (to fix the idea) asymptotic one to the other. We denote their arc length counted from their origins n and n' by t. Then, after a suitable selection of n and n', we have:

$$d(\gamma(t), \gamma'(t)) \leq \mathrm{nn}' e^{-t}, \qquad t \geq 0,$$

where d *means the Riemannian distance, and* nn' *is the arc-length of the horocycle.*

Proof:

Origins n and n' are selected on the same horocycle 1 (Figure A 20.19). Denote by m and m' the intersections of γ and γ' with another horocycle 2. Arcs nm and $\mathrm{n'm'}$ are equal, for 1 and 2 are parallel curves:

$$\mathrm{nm} = \mathrm{n'm'} = t .$$

Let us compute the arc mm' that belongs to 2. Horocycle 2 has the equation:

$$x = r \sin u, \quad y = r + r \cos u .$$

Thus, with obvious notations:

$$\mathrm{mm}' = \int_m^{m'} \frac{\sqrt{dx^2 + dy^2}}{y} = \int_m^{m'} \frac{du}{1 + \cos u} = tg \frac{u_{m'}}{2} - tg \frac{u_m}{2} .$$

Symmetrically, on horocycle 1:

$$\mathrm{nn}' = tg \frac{u_{n'}}{2} - tg \frac{u_n}{2} .$$

A straightforward computation with γ and γ' leads to:

$$t = \mathrm{nm} = \mathrm{Log}\left| tg \frac{u_n}{2} \right| - \mathrm{Log}\left| tg \frac{u_m}{2} \right| ,$$

$$t = \mathrm{n'm'} = \mathrm{Log}\left| tg \frac{u_{n'}}{2} \right| - \mathrm{Log}\left| tg \frac{u_{m'}}{2} \right| .$$

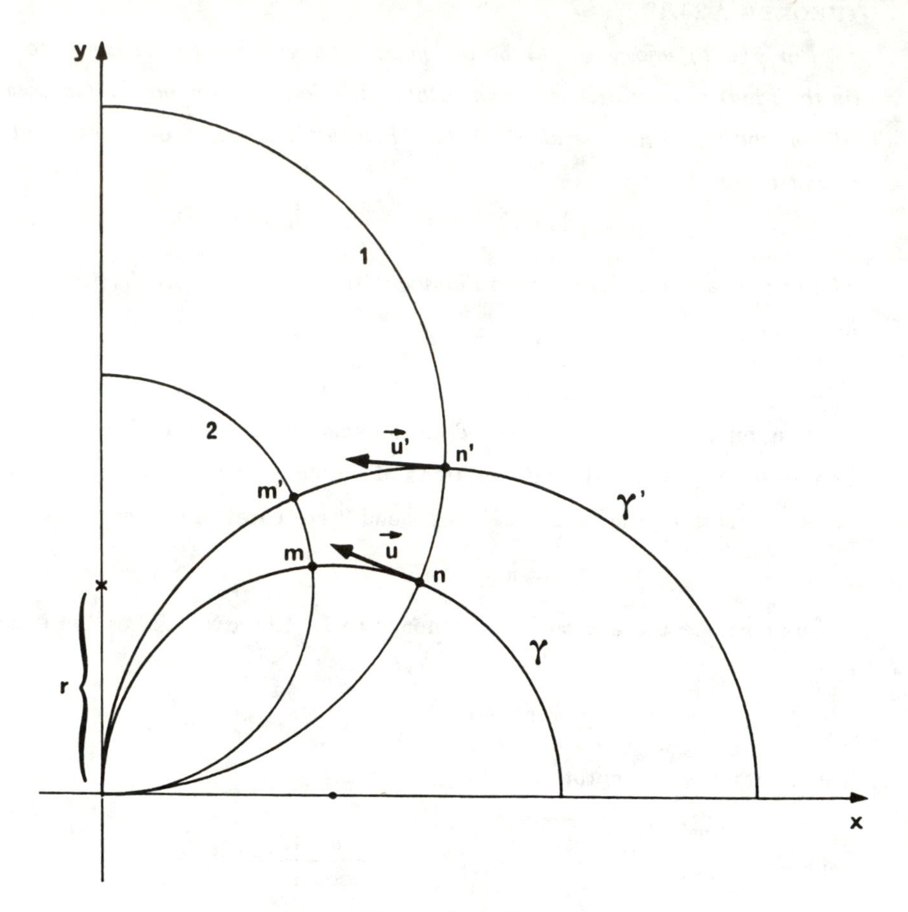

Figure A 20.19

Consequently:

$$e^t = \frac{tg\,\frac{u_n}{2}}{tg\,\frac{u_m}{2}} = \frac{tg\,\frac{u_{n'}}{2}}{tg\,\frac{u_{m'}}{2}} = \frac{tg\,\frac{u_{n'}}{2} - tg\,\frac{u_n}{2}}{tg\,\frac{u_{m'}}{2} - tg\,\frac{u_m}{2}} = \frac{\mathrm{nn}'}{\mathrm{mm}'}$$

$$\mathrm{mm}' = \mathrm{nn}' \cdot e^{-t} .$$

Theorem (A 20.18) follows from $d(m, m') \leq \mathrm{mm}'$.

GENERALIZATION A 20.20

The manifold V is the upper space $x_n > 0$ of R^n endowed with the metric:

$$ds^2 = \frac{(dx_1)^2 + \cdots + (dx_n)^2}{(x_n)^2} .$$

V is the Lobatchewsky space of constant curvature -1. The horocycles are $(n-1)$-dimensional manifolds, namely the planes x_n = constant and the Euclidean spheres of V which are tangent to the plane $x_n = 0$.

APPENDIX 21

PROOF OF THE

LOBATCHEWSKY-HADAMARD THEOREM

(See 14.3, Chapter 3)

§ A. Manifolds of Negative Curvature

First, let us recall some classical properties of Riemannian manifolds of negative curvature.

THEOREM A 21.1

Let V be a complete, simply connected Riemannian manifold of negative curvature. Then:

(1) *There exists one, and only one, geodesic passing through two distinct given points;*

(2) *V is diffeomorphic to the Euclidean space;*

(3) *let ABC be a geodesic triangle whose angles are A, B, C and whose sides are a, b, c. Then:*

$$a^2 + b^2 - 2ab \cdot \cos C \leq c^2 \quad .$$

Proof will be found in S. Helgason [1].

A direct consequence is the following corollary:

COROLLARY A 21.2

Under the above assumptions, Riemannian spheres of V are convex, that is, a geodesic has at most two common points with a sphere.

§ B. Asymptotes to a Given Geodesic

As usual, $\gamma(x, \mathbf{u}, t) = \gamma(t) = \gamma$ denotes a geodesic emanating from x, with initial velocity-vector $\mathbf{u}$ and arc length t. The point of γ corresponding to t is denoted also by $\gamma(t)$. The Riemannian distance of two points a and b is denoted by $|a, b|$. Denote a complete, simply connected Riemannian manifold of negative curvature by V.

THEOREM A 21.3

Let v' be a point of V. The geodesic joining v' to a point $\gamma(t) \in \gamma$ converges to a limit as $t \to +\infty$ (resp. $t \to -\infty$). This limit is a geodesic.

Proof: (See Figure A 21.4.)

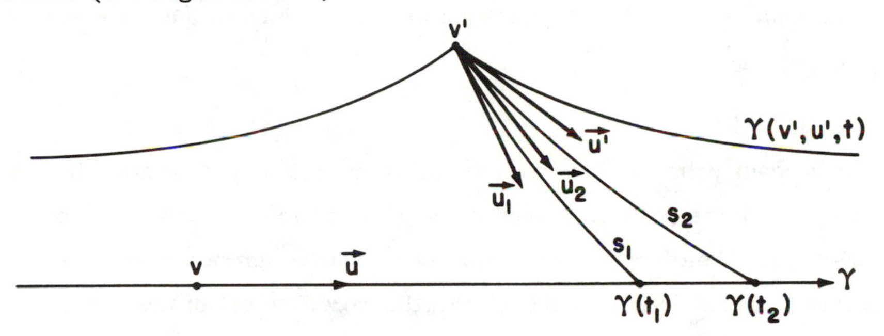

Figure A 21.4

The points v' and $\gamma(t_1)$ define one, and only one, geodesic $\gamma(v', \mathbf{u}_1, t)$. We set $s_1 = |v, \gamma(t_1)|$. Take $t_2 > t_1$ and apply relation (3) of Theorem (A 21.1) to the geodesic triangle v', $\gamma(t_1)$, $\gamma(t_2)$. With obvious notations we have:

$$(s_2)^2 - (s_1)^2 - (t_2 - t_1)^2 \leq 2 s_1 s_2 \cdot \cos(\mathbf{u}_1, \mathbf{u}_2) .$$

On the other hand, the triangular inequality applied to v, v' $\gamma(t_1)$ gives:

$$t_1 - |v, v'| \leq s_1 \leq t_1 + |v, v'| ,$$

whence:

$$s_1 = t_1 + O(1), \qquad t_1 \to +\infty .$$

Similarly:

$$s_2 = t_2 + O(1), \quad t_2 \to +\infty .$$

We deduce:

$$\lim_{t_1, t_2 \to +\infty} \cos(\mathbf{u}_1, \mathbf{u}_2) = 1,$$

that is to say:

$$\lim_{t_1, t_2 \to +\infty} \widehat{(\mathbf{u}_1, \mathbf{u}_2)} = 0 .$$

Thus, according to Cauchy, $\mathbf{u}_1$ converges to a limit $\mathbf{u}'$ as $t_1 \to +\infty$.

The geodesic $\gamma(v', \mathbf{u}', t)$ is the limit position of $\gamma(v', \mathbf{u}_1, t)$, for the exponential mapping $\mathrm{Exp}_{v'}$ is continuous. *$\gamma(v', \mathbf{u}', t)$ is called a positive asymptote to γ.* Negative asymptotes are defined in the same way ($t_1 \to -\infty$).

REMARK A 21.5

It is readily proved that the positive asymptote to γ emanating from a given point of the positive asymptote $\gamma(v', \mathbf{u}', t)$ is nothing but γ (geometrically). Therefore, we may speak of a positive asymptote to γ without referring to a definite point v'. Furthermore, the set of the positive asymptotes to γ is a $(\dim V - 1)$- parameter family of geodesics.

§ C. The Horospheres[1] of V

The Riemannian manifold V is again complete, simply connected, and of negative curvature. Let $\gamma(v, \mathbf{u}, t) = \gamma(t)$ be a geodesic and v' an arbitrary point of V.

LEMMA A 21.6

$$|v', \gamma(t)| - |v, \gamma(t)| \equiv \phi(t)$$

converges to a finite limit $L(v' : \gamma, v)$ as $t \to +\infty$, and this limit is a C^1-differentiable function of v' and v.

---

[1] See A. Grant [1].

Proof:

Take $t_2 > t_1$. The triangular inequality applied to v', $\gamma(t_1)$, $\gamma(t_2)$ gives:

$$\phi(t_2) = |v', \gamma(t_2)| - |v, \gamma(t_2)| \leq |v', \gamma(t_1)| + |\gamma(t_1), \gamma(t_2)| - |v, \gamma(t_2)|$$

$$= |v', \gamma(t_1)| - |v, \gamma(t_1)| = \phi(t_1) .$$

Therefore, $\phi(t)$ decreases monotonically. On the other hand, $\phi(t)$ is bounded, for the triangular inequality applied to $v, v', \gamma(t)$ gives:

$$|\phi(t)| = ||v', \gamma(t)| - |v, \gamma(t)|| \leqq |v, v'| .$$

This proves the existence of:

$$\lim_{t \to +\infty} \phi(t) = L(v'; \gamma, v) .$$

The second assertion follows from the inequality:

$$|[|v_1', \gamma(t)| - |v_1, \gamma(t)|] - [|v', \gamma(t)| - |v, \gamma(t)|]| \leq |v', v_1'| + |v, v_1| ,$$

that is

$$|L(v_1'; \gamma, v_1) - L(v'; \gamma, v)| \leq |v', v_1'| + |v, v_1| .$$

Obviously:

$$L(v'; \gamma, v) - L(v'; \gamma, v_1) = \overline{vv_1} , \tag{A 21.7}$$

where $\overline{vv_1}$ is the algebraic measure of vv_1 on the oriented geodesic γ.

DEFINITION A 21.8

The locus of the points x *for which* $L(x; \gamma, O) = 0$ *is called the positive horosphere through* O *of* γ *and will be denoted by* $H^+(\gamma, O)$.

According to Lemma (A 21.7), $H^+(\gamma, O)$ is a C^1-differentiable submanifold of dimension $(\dim V - 1)$. Let v_1 be an arbitrary point of γ. Relation (A 21.7) shows that $H^+(\gamma, O)$ has equation:

$$L(x; \gamma, v_1) = \overline{Ov_1} .$$

Now we obtain the horospheres as spheres with center at infinity and radius infinite. The Riemannian sphere, the center of which is a and passing through b, will be denoted by $\Sigma(a, b)$.

Lemma A 21.9

$\Sigma(\gamma(t), O)$ converges to $H^{+}(\gamma, O)$ as $t \to +\infty$.

Proof:

Let x be a point of $H^{+}(\gamma, O)$, we have:

$$\phi(t) \equiv |x, \gamma(t)| - |O, \gamma(t)| \to 0 \text{ as } t \to +\infty .$$

On the other hand, $\phi(t) \geq 0$. Therefore, $\Sigma(\gamma(t), O)$ intersects the geodesic segment $x\gamma(t)$ at a point $b(t)$ (see Figure A 21.10).

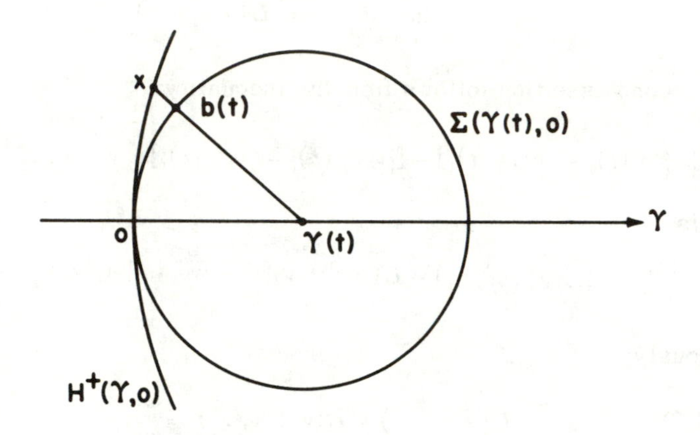

Figure A 21.10

We have:

$$|x, b(t)| = |x, \gamma(t)| - |\gamma(t), b(t)| = |x, \gamma(t)| - |O, \gamma(t)| \to 0 \text{ as } t \to +\infty.$$

This means that every point of $H^{+}(\gamma, O)$ is a limit point of the spheres $\Sigma(\gamma(t), O)$ as $t \to +\infty$. Conversely, we prove that such a limit point belongs to $H^{+}(\gamma, O)$. Let $b(t)$ be a point of $\Sigma(\gamma(t), O)$ and $x = \lim_{t \to +\infty} b(t)$. The triangular inequality gives:

$$\big|\,|x, \gamma(t)| - |O, \gamma(t)|\,\big| \leq \big|\,|x, \gamma(t)| - |b(t), \gamma(t)|\,\big| + \big|\,|b(t), \gamma(t)| - |O, \gamma(t)|\,\big|$$

$$= |x, b(t)| \to 0 \text{ as } t \to +\infty .$$

Therefore: $L(x;\ \gamma, O) = 0$, that is,

$$x \in H^+(\gamma, O)\ .$$

COROLLARY A 21.11

Horospheres are convex, and strictly convex if the curvature of V is bounded from above by a negative constant.

Proof:

$H^+(\gamma, O)$ is the limit of the balls passing through O and the center of which goes to infinity along γ, and these balls are convex (see A 21.2).

LEMMA A 21.12

Let $H^+(\gamma, O)$ and $H^+(\gamma, O')$ be two horospheres of γ. If $a \in H^+(\gamma, O)$ and $a' \in H^+(\gamma, O')$, then $|a, a'| \geq |O, O'|$.

Proof:

Assume $|a, a'| < |O, O'|$. From (A 21.9) we conclude that to each t corresponds a point $a(t) \in \Sigma(\gamma(t), O)$ such that:

$$\lim_{t \to +\infty} a(t) = a,$$

and a point $a'(t) \in \Sigma(\gamma(t), O')$ such that:

$$\lim_{t \to +\infty} a'(t) = a'\ .$$

Thus, for t large enough, we have:

$$|a(t), a'(t)| < |O, O'|\ .$$

To fix the ideas assume that the point O' lies between the points O and $\gamma(t)$. We obtain the following contradiction:

$$|a(t), \gamma(t)| \leq |a(t), a'(t)| + |a'(t), \gamma(t)| < |O, O'| + |a'(t), \gamma(t)|$$

$$= |O, \gamma(t)| = |a(t), \gamma(t)|\ . \qquad \text{(Q.E.D.)}$$

LEMMA A 21.13

Two positive horospheres $H^+(\gamma, O)$ and $H^+(\gamma, O')$ cut off an arc of length $|O, O'|$ on every positive asymptote to γ.

Proof:

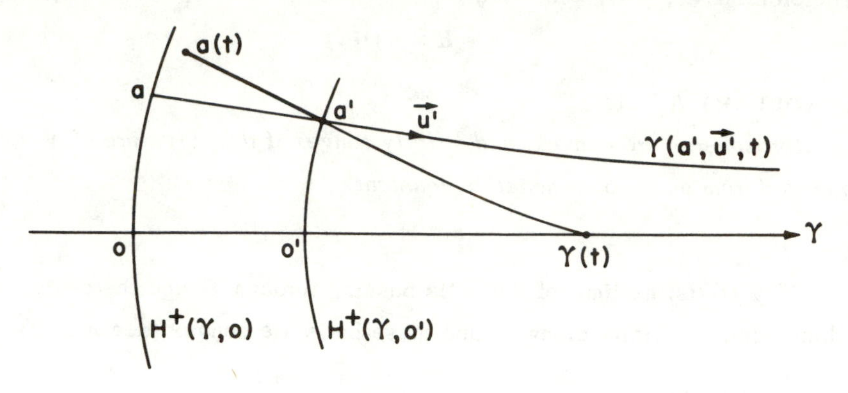

Figure A 21.14

Let $\gamma(a', \mathbf{u}, t)$ be a positive asymptote to γ that intersects $H^+(\gamma, O')$ at a'. The points $\gamma(t)$ and a' define a geodesic on which we select a point $a(t)$ such that $|a(t), a'| = -L(a'; \gamma, O) = |O, O'|$ and a' lies between $a(t)$ and $\gamma(t)$ (see Figure A 21.14). Since the exponential mapping $\mathrm{Exp}_{a'}$ is continuous, we obtain:

$$\lim_{t\to+\infty} a(t) = a \in \gamma(a', \mathbf{u}', t) \quad \text{and} \quad |a, a'| = -L(a'; \gamma, O) .$$

We deduce:

$$\begin{aligned} \big|\,|a, \gamma(t)| - |O, \gamma(t)|\,\big| &\leq \big|\,|a, a(t)| + |a(t), \gamma(t)| - |O, \gamma(t)|\,\big| \\ &= \big|\,|a, a(t)| + |a', \gamma(t)| - |O', \gamma(t)|\,\big| \to 0 \quad \text{as} \quad t \to +\infty . \end{aligned}$$

Thus, $a \in H^+(\gamma, O)$.

(Q. E. D.)

THEOREM A 21.15

The positive asymptotes to γ *are the orthogonal trajectories of the positive horospheres of* γ.

Proof:

Direct consequence of (A 21.12) and (A 21.13).

Finally, observe that negative horospheres $H^-(\gamma, O)$ can be defined as above from the negative asymptotes $(t \to -\infty)$.

§ D. The Horospheres of $T_1 V$

The unitary tangent bundle of V is denoted by $T_1 V$ and $p: T_1 V \to V$ is the canonical projection.

Let $\mathbf{u}$ be a point of $T_1 V$; $\mathbf{u}$ defines a geodesic $\gamma(p\mathbf{u}, \mathbf{u}, t) = \gamma(\mathbf{u}, t) = \gamma(t)$ *the lift of which, in* $T_1 V$, *is denoted again by* $\gamma(t)$. From § B we know there exist two horospheres $H^+(\gamma, p\mathbf{u}) = H^+(\mathbf{u})$ and $H^-(\gamma, p\mathbf{u}) = H^-(u)$ passing through $p\mathbf{u}$. The set of the unitary vectors orthogonal to $H^+(\mathbf{u})$ (resp. $H^-(\mathbf{u})$) along $H^+(\mathbf{u})$ (resp. $H^-(\mathbf{u})$) and oriented like $\mathbf{u}$ is a (dim $V-1$)-dimensional submanifold $\mathcal{H}^+(\mathbf{u})$ (resp. $\mathcal{H}^-(\mathbf{u})$) of $T_1 V$. *The* $\mathcal{H}$*'s are called the horospheres of* $T_1 V$.

THEOREM A 21.16

(1) *The* $\gamma(\mathbf{u}, t)$*'s and the* $\mathcal{H}^+(\mathbf{u})$*'s,* $\mathcal{H}^-(\mathbf{u})$*'s are the sheets of three foliations of* $T_1 V$.

(2) *At each point* $\mathbf{u} \in T_1 V$ *these foliations are transverse, that is:*

$$T(T_1 V)_{\mathbf{u}} = X^+_{\mathbf{u}} \oplus X^-_{\mathbf{u}} Z_{\mathbf{u}} ,$$

where $X^+_{\mathbf{u}}$ *(resp.* $X^-_{\mathbf{u}}$, $Z_{\mathbf{u}}$*) is the tangent space of* $\mathcal{H}^+(\mathbf{u})$ *(resp.* $\mathcal{H}^-(\mathbf{u})$, $\gamma(\mathbf{u}, t)$ *) at* $\mathbf{u}$.

(3) *These foliations are invariant under the geodesic flow* ϕ_t*:*

$$\phi_t \mathcal{H}^{\pm}(\mathbf{u}) = \mathcal{H}^{\pm}(\phi_t \mathbf{u}) ,$$

$$\phi_t \gamma(\mathbf{u}, t') = \gamma(\phi_t \mathbf{u}, t') .$$

Proof:

(1) Follows from the very construction of the sheets.

(2) Follows from the strict convexity of H^+ (resp. H^-, see A 21.11).

(3) Follows from Theorem (A 21.15).

The invariance of the foliations reduces the study of the differential ϕ_t^* to the study of its restriction to $\mathcal{H}^+(\mathbf{u})$ (resp. $\mathcal{H}^-(\mathbf{u})$) and $\gamma(\mathbf{u})$. Now, *we assume definitively that* V *is the universal covering* $\tilde{W}$ *of a compact Riemannian manifold* W *of negative curvature.* In particular, the curvature of V is bounded from above by a negative constant $-k^2$.

LEMMA A 21.17

Let $r_s(t)$ *be a one-parameter family* $(s > 0)$ *of numerical,* C^2*-differentiable functions. Assume that:*

$$\ddot{r}_s \geq k^2 \cdot r_s \qquad (k = \text{constant} > 0)$$

for every $s, t \geq 0$, *and* $r_s(0) > 0$, $r_s(s) = 0$. *Then:*

$$r_s(t) < r_s(0) \cdot \frac{\cosh[k(s-t)]}{\cosh[ks]}, \text{ for } 0 \leq t \leq s\ .$$

Assume, additionally, that:

$$\lim_{s \to +\infty} \dot{r}_s \Big|_{t=s} = 0\ .$$

Then, for s *large enough:*

$$|\dot{r}_s(t)| < k \cdot r_s(0) \cdot \frac{\sinh[k(s-t)]}{\cosh(ks)}\ , \text{ for } 0 \leq t \leq 4\ .$$

Proof:

The function:

$$l_s(t) = r_s(t) - \frac{r_s(0)}{\cosh(ks)} \cdot \cosh[k(s-t)]$$

verifies:

$$\ddot{l}_s(t) \geq k^2 \cdot l_s(t), \qquad l_s(0) = l_s(s) = 0\ .$$

Thus, l_s is concave between 0 and s and vanishes for $t = 0, s$, consequently $l_s(t) \leq 0$ for $0 \leq t \leq s$. This proves the first part. This proves also that $\dot{l}_s$ increases between 0 and s; thus: $\dot{l}_s(t) \leq \dot{l}_s(s)$ for $0 \leq t \leq s$. On the other hand,

$$\dot{l}_s(t) = \dot{r}_s(t) + k \cdot r_s(0) \cdot \frac{\sinh[k(s-t)]}{\cosh(ks)}\ ,$$

in particular,

$$\dot{l}_s(t) \leq \dot{l}_s(s) = \dot{r}_s(s) \to 0 \text{ as } s \to +\infty\ .$$

The second part follows easily.

THEOREM A 21.18

Let ϕ_t be the geodesic flow of T_1V. Then, for any positive number

$$\|\phi_t^* \xi\| \leq b \cdot e^{-kt} \cdot \|\xi\|, \ \|\phi_{-t}^* \xi\| \geq a \cdot e^{kt} \|\xi\| \ \text{ if } \xi \in X_{\mathbf{u}}^+ ,$$

$$\|\phi_t^* \xi\| \geq a \cdot e^{kt} \|\xi\|, \ \|\phi_{-t}^* \xi\| \leq b \cdot e^{-kt} \|\xi\| \ \text{ if } \xi \in X_{\mathbf{u}}^- .$$

The positive constants a and b are independent of t and ξ, and $\| \ \|$ denotes the length of a vector of T_1V equipped with its natural Riemannian metric.

Proof:

We prove the first inequality, the others can be proved in the same way.

Let $\gamma(O, \mathbf{u}, t) = \gamma(t) = \gamma$ be a geodesic of V, and let x be a point of $H^+(\gamma, O)$, close enough to O. There is a well-defined geodesic $\gamma_s(x, \mathbf{u}_s, t) = \gamma_s(t)$ passing through x and $\gamma(s) \in \gamma$. Our first purpose is to compute the Riemannian distance of $\dot{\gamma}(t)$ and $\dot{\gamma}_s(t)$, regarded as elements of T_1V. Let $r_s(t)$ be the Riemannian distance of their projections $\gamma(t)$ and $\gamma_s(t)$ on V. To compute $r_s(t)$ we consider a Jacobi field[2] $\psi(t)$ along γ, that is orthogonal to γ and vanishes for $t = s$. By definition:

$$< R(\dot{\gamma}, \psi)\dot{\gamma}, \psi > \; = \; - < \nabla^2 \psi, \psi > ,$$

where $R(\ , \ )$ is the curvature tensor and ∇ is the covariant derivative along γ. By definition the sectional curvature in the two-plane $(\dot{\gamma}, \psi)$ is:

$$\rho(\dot{\gamma}, \psi) = \frac{< R(\dot{\gamma}, \psi)\dot{\gamma}, \psi >}{\|\psi\|^2} .$$

We know that $\rho(\dot{\gamma}, \psi) \leq -k^2$, consequently:

$$< \nabla^2 \psi, \psi > \; \geq \; k^2 \cdot \|\psi\|^2 .$$

On the other hand,

$$< \nabla^2 \psi, \psi > \; = \; \nabla < \nabla \psi, \psi > - \|\nabla \psi\|^2 ,$$

---

[2] See J. Milnor [1].

$$\nabla < \nabla\psi, \psi > \; = \; \tfrac{1}{2}\nabla^2 \|\psi\|^2 \; = \; \tfrac{1}{2}\frac{d^2}{dt^2}\|\psi\|^2 \, ,$$

$$\|\nabla\psi\|^2 \geq \left(\frac{d}{dt}\|\psi\|\right)^2 \, .$$

Therefore, the length $l_s(t)$ of $\psi(t)$ verifies:

$$\tfrac{1}{2}(l_s^2)'' - (l_s')^2 \geq k^2 \cdot l_s^2 \, ,$$

that is

$$\ddot{l}_s \geq k^2 \cdot l_s, \text{ and } l_s(0) > 0, \quad l_s(s) = 0 \, .$$

Lemma (A 21.17) and the classical possibility to select the Jacobi field ψ such that:

$$r_s(t) = l_s(t) + 0(1)$$

if x is close enough to γ imply:

$$\text{(A 21.19)} \qquad r_s(t) < r_s(0) \cdot \frac{\cosh[k(s-t)]}{\cosh(ks)} \, , \quad \text{for } 0 \leq t \leq s \, .$$

Now it is readily seen that the angle of γ and γ_s at $\gamma(s)$ converges to zero as $s \to +\infty$. Thus, $\dot{r}_s(s) \to 0$ as $s \to +\infty$, and Lemma (A 21.17) implies again:

$$\text{(A 21.20)} \qquad |\dot{r}_s(t)| < k \cdot r_s(0) \frac{\sinh[k(s-t)]}{\sinh(ks)} \, , \quad \text{for } 0 \leq t \leq s \, .$$

As $s \to +\infty$, $\gamma_s(t)$ converges to a point $\gamma'(t)$ of the positive asymptote $\gamma'(x, u', t)$ to γ, and $\dot{\gamma}_s(t)$ converges to $\dot{\gamma}'(t)$. If $r(t)$ denotes the distance of $\gamma(t)$ to $\gamma'(t)$, then the inequalities (A 21.19) and (A 21.20) imply ($s \to +\infty$):

$$r(t) < r(0) \cdot e^{-kt} \, ,$$

$$|\dot{r}(t)| < kr(0)\, e^{-kt}, \quad \text{for } t > 0 \, .$$

Thus, the Riemannian distance of $\dot{\gamma}(t)$, $\dot{\gamma}'(t) \in T_1V$ verifies:

$$d(\dot{\gamma}(t), \dot{\gamma}'(t)) \leq r(0) \cdot \sqrt{1+k^2} \cdot e^{-kt}, \quad \text{for } t \geq 0.$$

We easily deduce the first inequality of Theorem (A 21.18).

Due to this theorem, the sheets $\mathcal{H}^+(u)$ (resp. $\mathcal{H}^-(u)$) are called the "contracting" (resp. "dilating") sheets of T_1V.

§ E. Proof of the Lobatchewsky-Hadamard theorem[3]

THEOREM A 21.21

Let W be a compact, connected Riemannian manifold of negative curvature, then the geodesic flow on T_1W is a C-flow.

Proof:

Let $V = \tilde{W}$ be the universal covering of W equipped with the inverse image of the Riemannian metric of W under the canonical projection π: $\tilde{W} \to W$. V satisfies the assumptions of preceding sections. Thus the geodesic flow on T_1V verifies the conditions of C-flows: condition (0) is trivially fulfilled; condition (1) follows from Theorem (A 21.16); condition (2) follows from Theorem (A 21.18). We finish the proof by proving that π is compatible with the three foliations of $V = \tilde{W}$ and $T_1\tilde{W}$. The first homotopy group $\pi_1(W)$ is isomorphic to a group of automorphisms of $\tilde{W}$, for W is connected. The group $\pi_1(W)$ acts also as a group of automorphisms of $T_1\tilde{W}$: if $u', u'' \in T_1\tilde{W}$ are congruent mod $\pi_1(W)$, then $\mathcal{H}^{\pm}(u')$ and $\mathcal{H}^{\pm}(u'')$ are themselves congruent mod $\pi_1(W)$.

REMARK A 21.22

The horospheres of a compact, n-dimensional manifold W are diffeomorphic to R^{n-1}. In fact, let us consider the horosphere $\mathcal{H}^+$. It is a

---

[3] See J. Hadamard [1]. Proofs of Sections B and C are mainly due to H. Busemann: *Metric Methods in Finsler Spaces and Geometry*, Ann. Math. Study, No. 8, Princeton University Press.

paracompact manifold. Let S be a compact subset of $\mathcal{H}^+$. Then $\phi_t S$ is covered by a disk D of $\phi_t \mathcal{H}^+$ (take t large enough). The counterimage $\phi_t^{-1} D$ is a disk which covers S in $\mathcal{H}^+$. Therefore, $\mathcal{H}^+$ is diffeomorphic to $\mathbf{R}^{n-1}$, according to the following lemma of Brown (*Proc. Amer. Math. Soc.*, *12* (1961), 812-814) and Stallings (*Proc. Cambridge Philos. Soc.*, *58* (1962), 481-488): Let M be a paracompact manifold such that every compact subset is contained in an open set diffeomorphic to Euclidean space. Then M itself is diffeomorphic to Euclidean space.

This result does not hold for noncompact manifolds. Consider the space $\{(x, y) \mid y > 0,\ x \,(\mathrm{mod}\ 1)\}$ endowed with the metric:

$$ds^2 = \frac{dx^2 + dy^2}{y^2}$$

The Gaussian curvature is equal to -1 and the universal covering space is the Lobatchewsky plane (see Appendix 20). The curve $y = 1$ is an horocycle homeomorphic to S^1.

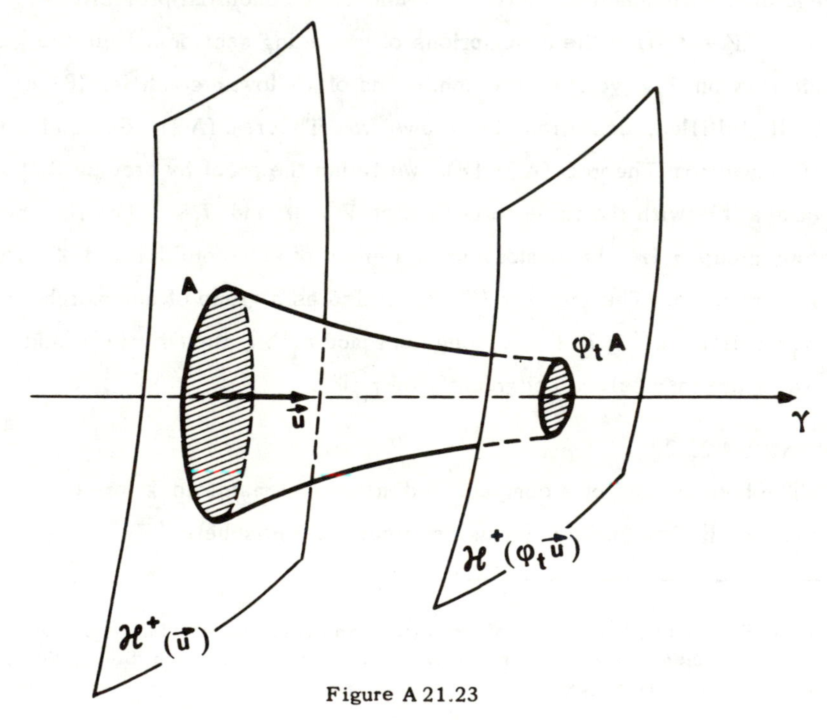

Figure A 21.23

APPENDIX 22

PROOF OF THE SINAI THEOREM

(See Section 15, Chapter 3)

Let (M, ϕ) be a C-diffeomorphism and X_m (resp. Y_m) the k-dimensional dilating space at m (resp. the l-dimensional contracting space). *A Riemannian metric is definitively selected on M.* Thus, X_m and Y_m are Euclidean subspaces of TM_m.

THE METRIC SPACE OF THE FIELDS OF TANGENT k-PLANES

A 22.1

The tangent space TM_m is the direct sum $X_m \oplus Y_m$. Therefore, the equation of a k-plane $U_m \subset TM_m$, transverse to Y_m, is:

$$y = P(U_m)x,$$

where $x \in X_m$, $y \in Y_m$, and $P(U_m): X_m \to Y_m$ is a linear mapping. We define a metric in accordance with the norm of the linear mappings $P(U)$: if U_m and U'_m are two k-planes of TM_m, then we set:

$$|U_m - U'_m| = \|P(U_m) - P(U'_m)\| = \sup_{x \in X_m, |x| < 1} |P(U_m)x - P(U'_m)x| .$$

We turn the set K of the fields of the tangent k-planes transverse to Y_m into a metric space by setting:

$$|U - U'| = \sup_{m \in M} |U_m - U'_m|, \qquad U, U' \in K.$$

The distance of inequality (15.3) of Section 15 has to be understood in this sense. Since M is compact, then K is a compact and complete metric space.

LEMMA A 22.2

Let R_1 and R_2 be two n-dimensional Euclidean spaces ($n = \dim M$). Assume that R_i ($i = 1,2$) is the direct sum of two subspaces X_i and Y_i:

$$R_i = X_i \oplus Y_i, \quad \dim X_i = k, \quad \dim Y_i = l:$$

Let $A\colon R_1 \to R_2$ be a linear mapping such that:

$$AX_1 = X_2, \quad AY_1 = Y_2,$$

$$\text{(A 22.3)} \qquad \begin{cases} \{ |Ax| \geq \mu |x| & \text{for } x \in X_1 \\ \{ |Ay| \leq \sigma |x| & \text{for } y \in Y_1 \end{cases},$$

where μ and σ are constants.

Let us denote by $\mathfrak{A}$ the operator induced by A, which makes correspond to the k-planes of R_1 the k-planes of R_2. If U and U' are transverse to Y_1, then:

$$|\mathfrak{A}U - \mathfrak{A}U'| \leq \mu^{-1}\sigma |U - U'| .$$

Proof:

By definition:

$$|\mathfrak{A}U - \mathfrak{A}U'| = \sup_{\substack{|x|<1 \\ x\in X_2}} |P(\mathfrak{A}U)x - P(\mathfrak{A}U')x|$$

$$= \sup_{\substack{|x|<1 \\ x\in X_2}} |A[P(U)A^{-1}x] - A[P(U')A^{-1}x]| .$$

According to (A 22.3), $|x| < 1$ and $x \in X_2$ imply $A^{-1}x \in X_1$ and $|A^{-1}x| \leq \mu^{-1}$. Therefore:

$$\sup_{\substack{|x|<1 \\ x\in X_2}} \leq \sup_{\substack{|z|<\mu^{-1} \\ x\in X_1}} ,$$

and

$$|\mathfrak{A}U - \mathfrak{A}U'| \leq \mu^{-1} \cdot \sup_{\substack{|z|<1 \\ z\in X_1}} |A[P(U) - P(U')]z| .$$

Since $[P(U) - P(U')]z \in Y_1$, (A 22.3) implies:

$$|A[P(U) - P(U')]z| \leq \sigma \cdot |[P(U) - P(U')]z|$$

$$\leq \sigma \cdot \sup_{\substack{|z|<1 \\ z \in X_1}} |[P(U) - P(U')]z| = \sigma \cdot |U - U'| .$$

Finally we have:

$$|AU - AU'| \leq \mu^{-1} \sigma |U - U'| .$$

(Q. E. D.)

INEQUALITY (15.3) OF SECTION 15. A 22.4

The mapping[1] ϕ^{**} *(or a positive integer power* ϕ^{**n} *is contracting in a neighborhood of the dilating field* X*:*

$$|\phi^{**} U_1 - \phi^{**} U_2| \leq \theta |U_1 - U_2|, \quad 0 < \theta < 1 ,$$

for

$$|X - U_1| < \delta, \quad |X - U_2| < \delta,$$

that is, for U_1 *and* U_2 *transverse to* Y.

Proof:

We apply the preceding lemma and we set:

$$R_2 = TM_m, \quad X_2 = X_m, \quad Y_2 = Y_m,$$

$$R_1 = TM_{\phi^{-n}(m)}, \ X_1 = X_{\phi^{-n}(m)}, \ Y_1 = Y_{\phi^{-n}(m)} .$$

The linear mapping A is the differential $(\phi^n)^*$ of ϕ^n. Since the dilating and the contracting fields X and Y are invariant under ϕ, $AX_1 = X_2$ and $AY_1 = Y_2$ are verified. Inequalities (A 22.3) are consequences of the axioms of C-systems:

$$\mu = a \cdot e^{\lambda n} , \qquad \sigma = b \cdot e^{-\lambda n} .$$

For n large enough we have

$$\theta = a^{-1} b \cdot e^{-2\lambda n} < 1 . \qquad \text{(Q. E. D.)}$$

---

[1] In fact ϕ^{**} is the extension of the mapping ϕ^{**} to the k-planes.

APPENDIX 23

SMALE CONSTRUCTION OF C-DIFFEOMORPHISMS

(See Section 12.3, Chapter 3)

Smale [3] has proved that there do exist nontoral C-diffeomorphisms. We give next an example of his construction.

The Space M.

Let G be the nilpotent Lie group of the 6×6 matrices:

$$g = \begin{pmatrix} 1 & x & y & & & \\ & 1 & z & & 0 & \\ & & 1 & & & \\ & & & 1 & X & Y \\ & 0 & & & 1 & Z \\ & & & & & 1 \end{pmatrix}$$

where $x, y, z, X, Y, Z \in R$. The group G is diffeomorphic to R^6.

Let us denote by $Q(\sqrt{3}) = \{p + q\sqrt{3} | p, q \in Z\}$ the number field of $\sqrt{3}$ adjoined to the rationals, and by $x = p + q\sqrt{3} \to \bar{x} = p - q\sqrt{3}$ the nontrivial Galois automorphism. We consider the subgroup Γ of G, the elements of which satisfy

$$x, y, z \in Q(\sqrt{3}),$$

$$X = \bar{x}, \quad Y = \bar{y}, \quad Z = \bar{z}.$$

It is readily proved that Γ is discrete and that the right coset space $M = \{g\Gamma\} = G/\Gamma$ is compact.

Of course, the first homotopy group of M is isomorphic to Γ and, consequently, is a nonabelian nilpotent group. Therefore, M is nontoral.

The Diffeomorphism ϕ: $M \to M$.

Let us identify an element $g \in G$ with (x, y, z, X, Y, Z). We define a mapping $\tilde{\phi}$: $G \to G$ by:

$$\tilde{\phi}(x, y, z, X, Y, Z) = (\lambda x, \mu y, \nu z, \bar{\lambda} X, \bar{\mu} Y, \bar{\nu} Z),$$

where:

$$\lambda = 2+\sqrt{3}, \quad \nu = (2-\sqrt{3})^2, \quad \mu = \lambda\nu = 2-\sqrt{3},$$

$\tilde{\phi}$ is an automorphism of G, because $\mu = \lambda\nu$. Therefore, $\tilde{\phi}\Gamma = \Gamma$, and $\tilde{\phi}$ defines a diffeomorphism ϕ of M by:

$$\phi: \; g\Gamma \to \tilde{\phi}(g)\Gamma .$$

(M, ϕ) is a C-Diffeomorphism.

An element of the Lie algebra TG_e of G is of the form

$$\begin{pmatrix} 0 & a & b & & & \\ & 0 & c & & 0 & \\ & & 0 & & & \\ & & & 0 & A & B \\ & 0 & & & 0 & C \\ & & & & & 0 \end{pmatrix}$$

The metric

$$ds^2 = da^2 + db^2 + dc^2 + dA^2 + dB^2 + dC^2$$

of TG_e defines a right invariant metric on G and, consequently, a Riemannian metric on $M = G/\Gamma$. The Lie algebra TG_e splits into the sum $X+Y$, where the elements of X (resp. Y) are of the form:

$$\begin{pmatrix} 0 & a & 0 & & & \\ & 0 & 0 & & & \\ & & 0 & & & \\ & & & 0 & 0 & B \\ & & & & 0 & C \\ & & & & & 0 \end{pmatrix} \quad \text{(resp.)} \quad \begin{pmatrix} 0 & 0 & b & & & \\ & 0 & c & & & \\ & & 0 & & & \\ & & & 0 & A & 0 \\ & & & & 0 & 0 \\ & & & & & 0 \end{pmatrix}$$

Next, by right translations, the splitting is imposed on the tangent space TG_g of every point g of G:

$$TG_g = \tilde{X}_g + \tilde{Y}_g .$$

Thus, the tangent space TM_m at $m \in M$ splits into:

$$TM_m = X_m + Y_m$$

It is easily checked that the linear tangent mapping $d\phi$ is dilating on X_m and contracting on Y_m.

APPENDIX 24

SMALE'S EXAMPLE

(See Section 16, Chapter 3)

Smale [2] proved the following theorem, which gives a negative answer to the "problem of structural stability": are the structurally stable diffeomorphisms dense in the C^1-topology?

THEOREM A 24.1

There exists a diffeomorphism $\psi: T^3 \to T^3$ such that no diffeomorphism ψ', C^1-close to ψ, is structurally stable.

We turn to the construction of ψ.

§ A. The Auxiliary Diffeomorphism ϕ

Let T^2 be the torus $\{(x, y) \bmod 1\}$. We define a diffeomorphism ϕ_1 of $T^2 \times \{z \mid -1 \le z \le 1\}$ onto itself by setting:

$$\phi_1: \begin{cases} \begin{pmatrix} x \\ y \end{pmatrix} \to \begin{pmatrix} 1 & 1 \\ 1 & 2 \end{pmatrix} \begin{pmatrix} x \\ y \end{pmatrix} \pmod 1 \\ z \to \tfrac{1}{2} z \end{cases}$$

Let $B_{1/2}$ be the ball (Figure A 24.2) of $T^2 \times \mathbf{R}$ with center $(0, 0, 2)$ and radius $1/2$:

$$x^2 + y^2 + (z-2)^2 \le \tfrac{1}{4}.$$

We define a diffeomorphism ϕ_1' of $B_{1/2}$ into $T^2 \times \{z \mid 0 \le z \le 3\}$ by setting:

$$\phi_1': \begin{cases} x \to \tfrac{1}{2} x \\ y \to \tfrac{1}{2} y \\ z \to 2z - 2. \end{cases}$$

Now, *the torus* T^3 *is* $T^2 \times S^1$, *where* S^1 *is* $[-3, 3]$ *with endpoints identified.*

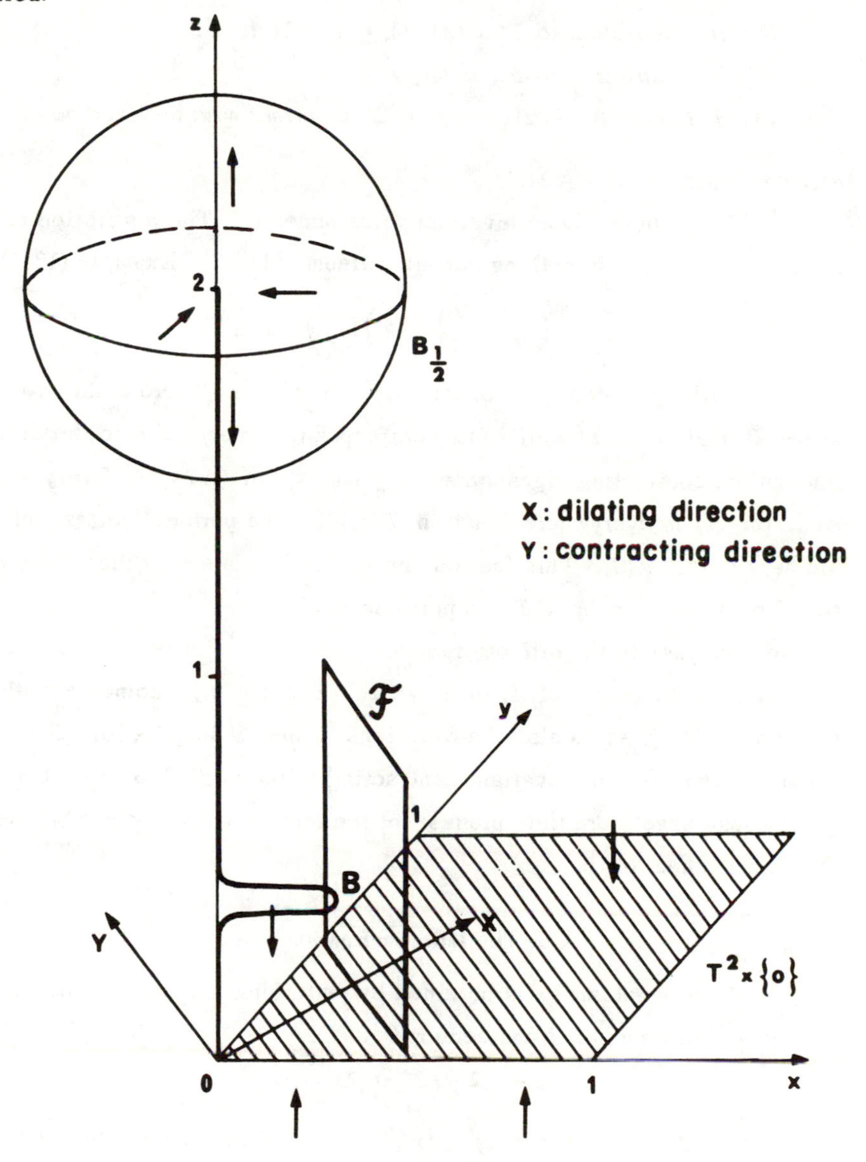

Figure A 24.2

The following lemma is easily proved.

LEMMA A 24.3

There exists a diffeomorphism $\phi: T^3 \to T^3$ *such that:*

(1) *its restriction to* $T^2 \times \{z \mid -1 \leq z \leq 1\}$ *is* ϕ_1;

(2) *its restriction to* $B_{1/2}$ *is* ϕ_1';

(3) ϕ *leaves* $\{(0, 0, z) \mid 0 \leq z \leq 2\}$ *invariant with no fixed point.*

PROPERTIES OF ϕ. A 24.4

$T^2 \times \{0\}$ is obviously an invariant torus under ϕ. The restriction of ϕ (or ϕ_1) to $T^2 \times \{0\}$ is nothing but the diffeomorphism of Example (13.1):

$$\begin{pmatrix} x \\ y \end{pmatrix} \to \begin{pmatrix} 1 & 1 \\ 1 & 2 \end{pmatrix} \begin{pmatrix} x \\ y \end{pmatrix} \pmod 1 . \tag{A 24.5}$$

Let us recall some properties of this diffeomorphism: There exist two foliations $\mathcal{X}$ and $\mathcal{Y}$ on $T^2 \times \{0\}$. They correspond, respectively, to the dilating and the contracting eigenspaces X_m and Y_m of (A 24.5). Every sheet of $\mathcal{X}$ (or $\mathcal{Y}$) is everywhere dense in $T^2 \times \{0\}$. The periodic points[1] of ϕ are dense in $T^2 \times \{0\}$. This fact can be proved by observing that every rational point $(p/q, p'/q) \in T^2$ is periodic.

Now we pass to the diffeomorphism $\phi: T^3 \to T^3$. It is easy to see that the periodic points of ϕ, in $T^2 \times \{z \mid -1 \leq z \leq 1\}$, coincide with those of (A 24.5), as do also the foliations $\mathcal{X}$ and $\mathcal{Y}$ in $T^2 \times \{0\}$. The foliation $\mathcal{Y}$ generates an invariant contracting foliation of $T^2 \times \{z \mid -1 \leq z \leq 1\}$, whose sheets are the "planes" of the form $Y \times \{z \mid -1 \leq z \leq 1\}$, where Y is some sheet of $\mathcal{Y}$.

§B. The Diffeomorphism ψ

The diffeomorphism ψ is obtained by perturbing ϕ. Let G_0 be the ball of T^3 with radius d and center $(0, 0, 3/4)$:

$$x^2 + y^2 + (z - 3/4)^2 \leq d^2 .$$

We set $G = \phi^{-1} G_0$, $\phi(x, y, z) = (x', y', z')$, and we observe that d can

---

[1] That is, the points $\xi \in T^2 \times \{0\}$ such that $\phi^N \xi = \xi$ for some nonvanishing integer N.

be chosen small enough for

$$\phi G \cap G = \emptyset \ .$$

We define our desired diffeomorphism ψ by setting:

$$\psi(x, y, z) = \begin{cases} \phi(x, y, z) = (x', y', z') \text{ outside } G \\ (x' + \eta\Phi(x, y, z),\ y',\ z') \text{ on } G, \end{cases}$$

where Φ is a nonnegative C^∞ function with compact support in G and nondegenerate maximum value $+1$ at $\phi^{-1}(0, 0, \tfrac{3}{4})$, and finally $\eta > 0$ is small enough so that ψ is a diffeomorphism.

PROPERTIES OF ψ. A 24.6

Now the curve $\psi\{(0, 0, z) | 0 \leq z \leq 2\}$ has a bump B (see Figure A 24.2). This bump lies in the region $T^2 \times \{z | -1 \leq z \leq 1\}$ where ψ coincides with $\phi = \phi_1$. This region is foliated into contracting planes (see A 24.4). Let $(x/a) + (y/b) = 1$ be the equation of such a contracting plane in the chart (x, y, z). Among these planes intersecting the bump B, we select the plane $\mathfrak{F}$, for which a is maximum (see Figure A 24.2). Either $\mathfrak{F}$ contains a periodic point of ψ, or it does not. In the first case, the bump is called periodic and in the second case, nonperiodic.

LEMMA A 24.7

ψ is not structurally stable.

This follows from two remarks:

(1) An arbitrarily small change of η in the definition of ψ gives rise to a diffeomorphism ψ'' arbitrarily C^1-close to ψ and similar to ψ. The density of the periodic points (see A 24.4) implies that we can suppose the bump of ψ is periodic and the bump of ψ'' is nonperiodic, and vice versa.

(2) If ψ and ψ'' are in the opposite cases there is no homeomorphism $h: T^3 \to T^3$ close to the identity such that $\psi'' \cdot h = h \cdot \psi$. In fact, h establishes a one-to-one correspondence between bumps, contracting planes, and periodic points of ψ and ψ''.

LEMMA A 24.8

Every diffeomorphism ψ', C^1*-close to* ψ, *possesses an invariant torus similar to* $T^2 \times \{0\}$, *a bump, a contracting sheet similar to* $\mathfrak{F}$, *and so on:* Complete proof of this lemma is announced by Smale [2]. Now Theorem (A 24.1) follows readily from Lemma (A 24.7) and (A 24.8).

APPENDIX 25

PROOF OF THE LEMMAS OF THE ANOSOV THEOREM

(See Section 16, Chapter 3)

Lemma A

Let (M, ϕ) be a C-diffeomorphism. *We select definitively a Riemannian metric on M.* Since M is compact, there exists a number $d > 0$ such that, whatever be the ball $B(p; d) \subset TM_p$ with radius d and center $p \in M$, the restriction

$$\mathrm{Exp}_p \Big|_{B(p\,;\,d)}$$

of the exponential mapping at p is a diffeomorphism. Let $\{\phi^n m \mid n \in \mathbf{Z}\}$ be an orbit of ϕ. A chart of a neighborhood of this orbit is (B, ψ^{-1}), where B is the sum of the balls

$$B_n = B(\phi^n m, d) \subset TM_{\phi^n m}$$

and the restriction $\psi|_{B_n}$ is $\mathrm{Exp}_{\phi^n m}$. Let us denote by X_n the dilating k-space $X(\phi^n m)$ of $TM_{\phi^n m}$, and by Y_n the contracting l-space $Y(\phi^n m)$. The invariant dilating and contracting foliations $\mathcal{X}$ and $\mathcal{Y}$ of M induce new foliations on B:

$$\mathcal{X}_1 = \psi^{-1}\mathcal{X}, \quad \mathcal{Y}_1 = \psi^{-1}\mathcal{Y}.$$

Finally, ϕ induces a mapping:

$$\phi_1 : B \to B, \quad \phi_1 = \psi^{-1}\phi\psi,$$

such that the restriction $\phi_1|_{B_n}$ maps B_n into B_{n+1}, with obvious restrictions concerning the range of ϕ_1. Assume d small enough, then the sheets of the foliations $\mathcal{X}_1$ and $\mathcal{Y}_1$ can be regarded as sheets of the Euclidean space $X_n \oplus Y_n$ of origin $0 = \phi^n m$, in which their equations are, respectively:

$$y = y(0) + f_n(x, y(0)) \quad \text{and} \quad x = x(0) + g_n(y, x(0))$$

where $x \in X_n$, $y \in Y_n$ and where f_n, g_n, and their first derivatives can be made arbitrarily small by a suitable choice of d.

Consider the sheet a_n of $\mathcal{Y}_n$ which passes through the center O of B_n:

$$x = g_n(y, 0) .$$

The mapping:

$$\mathcal{C}: \{Y_n \mid n \in \mathbf{Z}\} \to \{a_n \mid n \in \mathbf{Z}\} ,$$

whose restriction $\mathcal{C}|_{Y_n}: Y_n \to a_n$ is defined for $y \in Y_n \cap B_n$ by

$$\mathcal{C}y = (g_n(y, 0), y) ,$$

is a diffeomorphism. Therefore, $y \in Y_n$ can be regarded as coordinates on a_n.

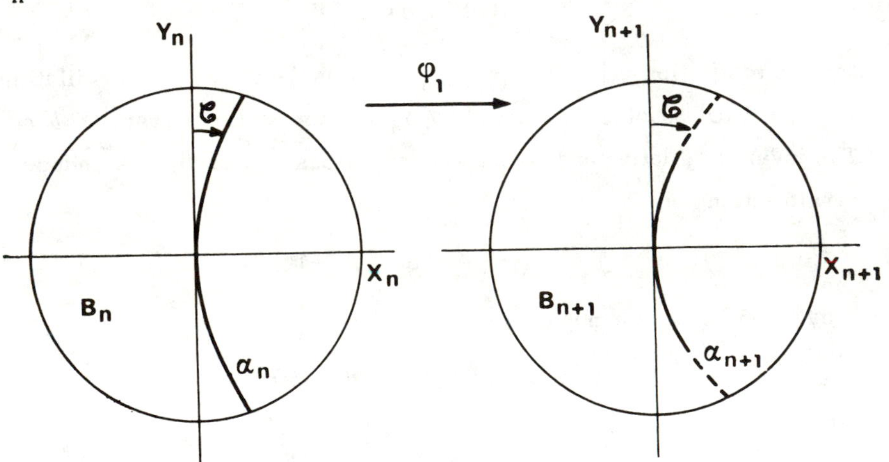

Figure A 25.1

The diffeomorphism ϕ_1 maps a_n into a_{n+1} (see Figure A 25.1). In the coordinates y, this defines a mapping ϕ_2:

$$\phi_2 = \mathcal{C}^{-1}\phi_1\mathcal{C},$$

and

$$\phi_2: \{Y_n \mid n \in Z\} \to \{Y_n \mid n \in Z\}, \quad \phi_2\big|_{Y_n}: Y_n \to Y_{n+1}.$$

Assertion A 25.2

It follows from the very definition of C-systems that the restriction

$$\phi_2\big|_{Y_n}: Y_n \to Y_{n+1}$$

is contracting:

(A 25.3) $$\|\phi_2 y\| < \theta\|y\|, \quad 0 < \theta < 1, \quad \text{for any } y \in Y_n \cap B_n,$$

where θ is a constant.

Remark A 25.4

To be precise, (A 25.3) holds for a certain iteration ϕ_2^ν of ϕ_2: we must "kill" the constant b in the definition of C-systems. For simplicity, we assume that (A 25.3) already holds for $\nu = 1$. Now let ϕ' be a diffeomorphism C^2-close to ϕ. Then ϕ' is a C-diffeomorphism (Sinai theorem, Section 15) and the foliations $\mathfrak{X}_1' = \psi^{-1}\mathfrak{X}$, $\mathfrak{Y}_1' = \psi^{-1}\mathfrak{Y}'$ and the mapping ϕ_1' induced by ϕ':

$$\phi_1': B \to B, \quad \phi_1' = \psi^{-1}\phi'\psi, \quad \phi_1'\big|_{B_n}: B_n \to B_{n+1},$$

are defined as above. If ϕ' is C^2-close enough to ϕ, then the sheets of $\mathfrak{X}_1'$ are close to those of $\mathfrak{X}_1$ and transverse to the sheet a_n. Therefore, there exists a projection Π:

$$\Pi: B \to \{a_n \mid n \in Z\}, \quad \Pi\big|_{B_n}: B_n \to a_n,$$

which makes correspond to each point $a \in B_n$, the intersection Πa of a_n with the sheet of $\mathfrak{X}_1'$ passing through a (see Figure A 25.5).

Now let us consider the mapping (Figure A 25.5):

$$\phi_2' = \mathcal{C}^{-1}\Pi\phi_1'\mathcal{C},$$

$$\phi_2' : \{Y_n \mid n \in \mathbf{Z}\} \to \{Y_n \mid n \in \mathbf{Z}\}, \quad \phi_2' \big|_{Y_n} : Y_n \to Y_{n+1} \, .$$

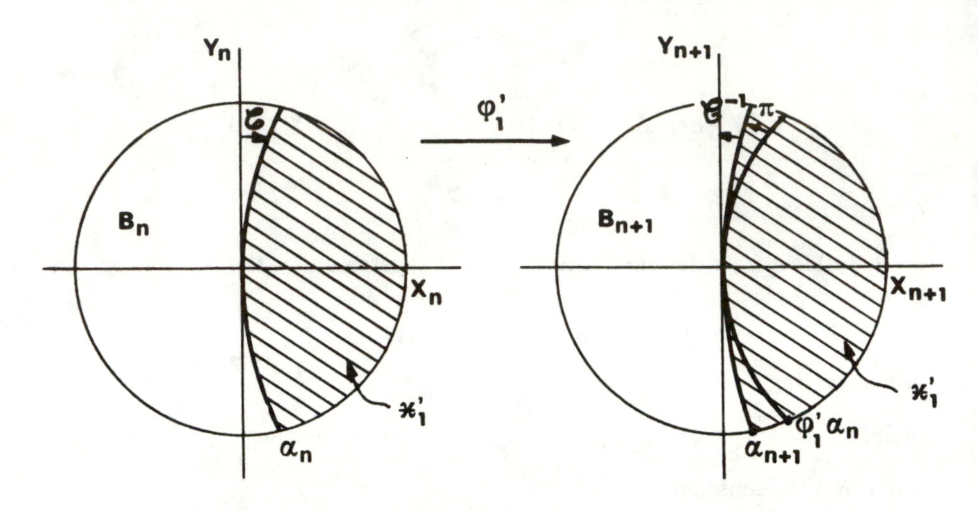

Figure A 25.5

ASSERTION A 25.6

If ϕ' is C^2-close enough to ϕ, then ϕ_2' is C-close to ϕ_2: to any æ > 0 corresponds a positive δ such that $\|\phi' - \phi\|_{c^2} < \delta$ implies:

(A 25.7) $$\|\phi_2' y - \phi_2 y\| < æ \text{ for any } y \in Y_n \cap B_n,$$

where $\| \ \|_{c^2}$ is the C^2-norm.

Proof:

ϕ_1' is close to ϕ_1, Π: $\phi_1' a_n \to a_{n+1}$ is small for $\phi_1' a_n \approx \phi_1 a_n = a_{n+1}$, and the sheets of $\mathfrak{X}_1'$ are transverse to a_{n+1}. Now denote the sheet of $\mathfrak{X}_1'$ passing through the center $0 = m$ of B_0 by β.

LEMMA A. A 25.8

If ϕ' is C^2-close enough to ϕ, then the sheet $\phi'^n \beta$ is close to $\phi^n m$ for any $n \geq 0$. To be precise:

(A 25.9) $$\|\phi_2'^n m\| < \frac{æ}{1-\theta} \, ,$$

where θ is defined at (A 25.3).

Proof:

According to (A 25.6), given $æ > 0$ there exists $\delta > 0$ such that $\|\phi' - \phi\|_{c^2} < \delta$ implies:

$$\|\phi_2' y - \phi_2 y\| < æ .$$

From (A 25.3) and (A 25.7) we deduce:

$$\|\phi_2' y\| < \theta \|y\| + æ .$$

Set $\varepsilon = æ/(1-\theta)$; therefore, if $\|y\| < \varepsilon$, then $\|\phi_2' y\| < \varepsilon$ and $\|\phi_2'^2 y\| < \varepsilon$, and so on. But since $\|m\| < \varepsilon$, inequality (A 25.9) is proved.

REMARK A 25.10

From (A 25.3) and (A 25.7) one deduces also that $\|y\| < c$ implies:

$$\|\phi_2'^n y\| \leq c \text{ if } c \geq \frac{æ}{1-\theta} .$$

In fact $\|y\| \leq c$ and $c \geq æ/(1-\theta)$ imply:

$$\|\phi_2' y\| < \theta \cdot \|y\| + æ < \theta c + æ < c .$$

(Q. E. D.)

Lemma B

Now we consider the sheet γ_n of $\mathfrak{X}$ which passes through $\phi^n m$. Let $\gamma_n^1 \subset B_n$ be the corresponding sheet of $\mathfrak{X}_1$:

$$\gamma_n^1 = \psi^{-1} \gamma_n .$$

The equation of γ_n^1 is $y = f_n(x, 0)$. We define a mapping $\mathfrak{D}$, similar to $\mathfrak{C}$, by setting:

$$\mathfrak{D}x = (x, f_n(x, 0)) \text{ for } x \in X_n \cap B_n ,$$

$$\mathfrak{D} : \{X_n \mid n \in Z\} \to \{\gamma_n^1 \mid n \in Z\}, \quad \mathfrak{D}|_{X_n} : X_n \to \gamma_n^1 .$$

$\mathfrak{D}$ is a diffeomorphism and $x \in X_n$ can be regarded as coordinates on γ_n^1. The diffeomorphism ϕ_1 maps γ_n^1 into γ_{n+1}^1. In the coordinates x, this defines a mapping:

$$\phi_3 = \mathfrak{D}^{-1} \phi_1 \mathfrak{D} : \{X_n \mid n \in Z\} \to \{X_n \mid n \in Z\}, \quad \phi_3|_{X_n} : X_n \to X_{n+1} .$$

Obviously $\phi_3(0) = 0$.

ASSERTION A 25.11

It follows from the very definition of C-systems that $\phi_3|_{X_n}: X_n \to X_{n+1}$ is dilating:

(A 25.12) $$\|\phi_3 x_1 - \phi_3 x_2\| > \Theta \cdot \|x_1 - x_2\|,$$

for any $x_1, x_2 \in B_n \cap X_n$, where $\Theta > 1$.

REMARK A 25.13

In fact, (A 25.12) holds for a certain iteration of ϕ_3. For simplicity we assume that (A 25.12) already holds for ϕ_3. Now let ϕ' be a diffeomorphism C^2-close to ϕ. Consider the sheet $\beta_n = \phi_1'^n \beta \subset B_n$ of the foliation $\mathfrak{X}_1' = \psi^{-1}\mathfrak{X}'$ which passes through $\mathcal{C}\phi_2'^n(0)$ $(n \geq 0)$. (See Figure A 25.14.) According to Lemma A, this sheet is close to the center O of B_n. Let $y = h_n(x)$, $(x \in X_n)$, be the equation of β_n. If ϕ' is close enough to ϕ, then we can choose $x \in X_n \cap B_n$ as local coordinates of β_n: the mapping

$$E: \{X_n | n \geq 0\} \to \{\beta_n | n \geq 0\}, \quad E|_{X_n}: X_n \to \beta_n,$$

which is defined by $x \to (x, h_n(x))$ for $x \in X_n \cap B_n$ is a diffeomorphism.

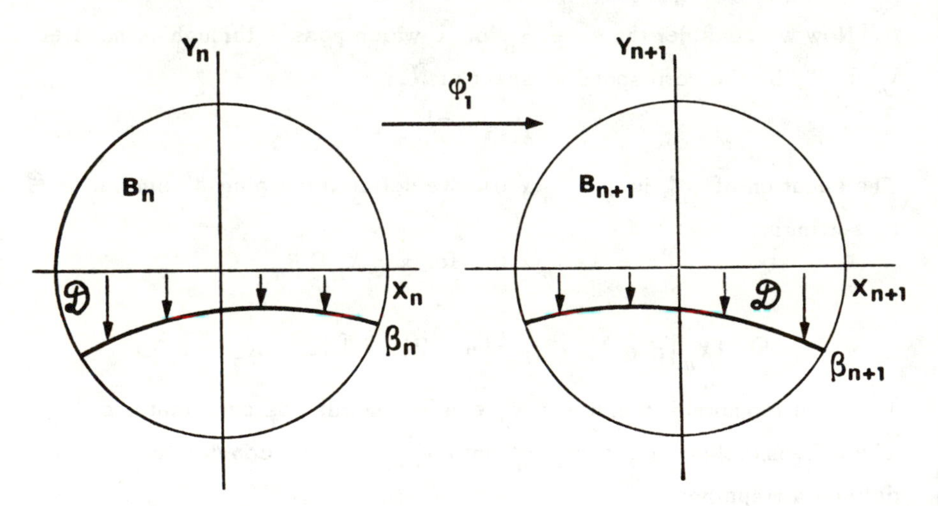

Figure A 25.14

From the very construction of the β_n's, we see that ϕ_1' maps β_n into β_{n+1}. Therefore, this defines a diffeomorphism:

$$\phi_3' = E^{-1}\phi_1' E : \{X_n | n \geq 0\} \to \{X_n | n \geq 0\}, \quad \phi_3'|_{X_n} : X_n \to X_{n+1}.$$

ASSERTION A 25.15

If ϕ and ϕ' are C^2-close enough, then ϕ_3 and ϕ_3' are C^1-close: To any $æ > 0$ corresponds a positive δ such that $\|\phi - \phi'\|_{c^2} < \delta$ implies:

$$\|\phi_3(x) - \phi_3'(x)\| < æ, \tag{A 25.16}$$

$$\|(\phi_3 - \phi_3')(x_1) - (\phi_3 - \phi_3')(x_2)\| < æ\|x_1 - x_2\|,$$

for any $x, x_1, x_2 \in X_n \cap B_n$, $n \geq 0$. This is a direct consequence of the construction of the γ_n, β_n (see Sinai theorem, Section 15), and of the fact that the β_n's are C^1-close to the γ_n's.

LEMMA B. A 25.17

If ϕ' is C^2-close enough to ϕ, then there exists a well-defined sheet $\delta \in \mathcal{Y}'$ such that $\phi'^n\delta$ is close to $\phi^n m$ for any $n \geq 0$. To be precise, there exists one and only one point $x_0 \in X_0$ such that $\|\phi_3'^n x_0\| < \varepsilon$ for any $n \geq 0$.

First we need a sublemma.

LEMMA A 25.18

Let R be a union of equi-dimensional Euclidean spaces R_n, $n \geq 0$. Let $T = K + L : R \to R$, $T|_{R_n} : R_n \to R_{n+1}$ be diffeomorphisms such that:

(1) $K(0) = 0$, $\|K(x) - K(y)\| > \Theta\|x - y\|$, $\Theta > 1$,

(2) $\|L\| \leq \varepsilon$, $\|L(x) - L(y)\| < \varepsilon\|x - y\|$, $\Theta - \varepsilon > 1$,

for any $x, y \in R$. Then, there exists one and only one point $x \in R_0$ such that the sequence $T^n x$ is bounded, and

$$\|T^n x\| \leq \frac{\varepsilon}{\Theta - 1} \quad \text{for any } n \geq 0. \tag{A 25.19}$$

Proof:

The mapping $T^{-1}|_{R_n}: R_n \to R_{n-1}$ $(n > 1)$ is obviously a diffeomorphism. On the other hand:

$$\|Tx - Ty\| = \|(Kx - Ky) + (Lx - Ly)\|$$

$$\geq \|Kx - Ky\| - \|Lx - Ly\| \geq (\Theta - \varepsilon)\|x - y\| ;$$

therefore:

(A 25.20) $$\|T^{-1}y - T^{-1}x\| < \frac{1}{\Theta - \varepsilon} \|y - x\| .$$

Let $b_n(c)$ be the ball $\|x\| \leq c$ of R_n. Since

$$\|Tx\| = \|Kx + Lx\| \geq \|Kx\| - \|Lx\| > \Theta \|x\| - \varepsilon ,$$

we have:

(A 25.21) $$Tb_n(c) \supset b_{n+1}(\Theta c - \varepsilon) .$$

Assume c large enough, that is:

(A 25.22) $$\Theta c - \varepsilon \geq c .$$

Then (A 25.21) implies $Tb_n(c) \supset b_{n+1}(c)$, therefore

$$T^{-1}b_{n+1}(c) \subset b_n(c) ,$$

consequently

$$T^{-1}b_1(c) \supset T^{-2}b_2(c) \supset \cdots \supset T^{-n}b_n(c) \supset \cdots .$$

But, according to (A 25.20), we have:

$$\text{diameter}\, T^{-n}b_n(c) \leq 2c(\Theta - \varepsilon)^{-n} \to 0 \text{ as } n \to +\infty .$$

Therefore $\bigcap_{n>0} T^{-n}b_n(C)$ reduces to a unique point $x \in b_0(c)$. This finishes the proof if one observes that $c = \varepsilon/(\Theta - 1)$ verifies (A 25.22).

Proof of Lemma B. A 25.23

According to (A 25.11) and (A 25.15), the mapping ϕ_3' verifies the conditions of the preceding lemma. It is sufficient to set $K = \phi_3$, $L = \phi_3' - \phi_3$, and change ε into æ in condition (2). If we take æ $= \varepsilon(\Theta - 1)$ in (A 25.19), we obtain $\|\phi_3'^{\,n} x_0\| < \varepsilon$. This proves Lemma B.

To summarize, we found a contracting sheet $\delta \in \mathcal{Y}'$ which remains close to the orbit $\phi^n m$ for $n \geq 0$ (in the sense of Lemma B). If ϕ' is close enough to ϕ, δ is close to $\phi^n m$, even for $n < 0$. To prove it, it is sufficient to apply Lemma A to ϕ^{-1}. The foliation $\mathcal{Y}'$ is the dilating foliation of ϕ'^{-1} and the sheet δ is close to m. Consequently, according to Remark (A 25.10), the sheets $\phi'^n\delta$, $(n < 0)$ stay in the neighborhood of the orbit $\phi^n m$ (in the sense of Lemma A):

$$\|\phi_2'^{-n} y\| \leq C .$$

Therefore, Lemmas A and B imply the following assertion:

ASSERTION A 25.24

If ϕ' is C^2-close enough to ϕ, then there exists a sheet $\tilde{\tilde{\delta}} \subset B_0$ of the foliation $\mathcal{Y}_1'$ such that the sheets $\phi_1'^n \tilde{\tilde{\delta}} \subset B_n$, $(-\infty < n < \infty)$, stay inside an ε-neighborhood of the center of B_n. Using the same argument for ϕ'^{-1}, we find a sheet $\tilde{\tilde{\beta}} \subset B_0$ of the foliation $\mathcal{X}_1'$ with similar properties. Since the sheets $\tilde{\tilde{\delta}}$ and $\tilde{\tilde{\beta}}$ are transverse in B_0, there exists one and only one point of intersection $z = \tilde{\tilde{\delta}} \cap \tilde{\tilde{\beta}}$ in an ε-neighborhood of the center of B_0. The desired homeomorphism k of the Anosov theorem is defined by setting $k(m) = \psi z$. One easily checks that all the preceding constructions depend continuously on m. This proves that k is an homeomorphism. The relation $\phi' k = k\phi$ is obvious, as is the fact that k is ε-close to the identity.

APPENDIX 26

INTEGRABLE SYSTEMS

(See Section 19, Chapter 4)

J. Liouville proved that if, in the system with n degrees of freedom:

$$\dot{p} = -\frac{\partial H}{\partial q}, \quad \dot{q} = \frac{\partial H}{\partial p}, \quad p = (p_1, \ldots, p_n), \quad q = (q_1, \ldots, q_n), \tag{A 26.1}$$

n first integrals in involution[1]

$$H = F_1, \ldots, F_n \ ; \ (F_i, F_j) = 0, \tag{A 26.2}$$

are known, then the system is integrable by quadratures.

Many examples of integrable problems of classical mechanics are known. In all these examples the integrals (A 26.2) can be found. It was pointed out long ago that, in these examples, the manifolds specified by the equations $F_i = f_i =$ constant turn out to be tori, and motion along them is quasi-periodic (compare with Example 1.2). We shall prove that such a situation is unavoidable in any problem admitting single-valued integrals (A 26.2). The proof is based on simple topological arguments.

THEOREM A 26.3

Assume that the equations $F_i = f_i =$ constant, $i = 1, \ldots, n$, *define an* n*-dimensional compact, connected manifold* $M = M_f$ *such that:*

(1) *at each point of* M *the gradients* grad F_i $(i = 1, \ldots, n)$ *are linearly independent;*

---

[1] Two functions $F(p, q)$ and $G(p, q)$ are in involution if their Poisson bracket vanishes identically:

$$(F, G) = \frac{\partial F}{\partial p}\frac{\partial G}{\partial q} - \frac{\partial F}{\partial q}\frac{\partial G}{\partial p} \equiv 0 .$$

(2) *a Jacobian* Det$|\partial I/\partial f|$, *which is defined below* (A 26.7) *does not vanish identically.*

Then:

(1) M *is an n-dimensional torus and the neighborhood of* M *is the direct product* $T^n \times \mathrm{R}^n$;

(2) *this neighborhood admits action-angle coordinates* (I, ϕ), $(I \in B^n \subset \mathrm{R}^n,\ \phi\ (\mathrm{mod}\ 2\pi) \in T^n)$, *such that the mapping* $I, \phi \to p, q$ *is canonical*[2] *and* $F_i = F_i(I)$.

Thus, Equations (A 26.1) may be written:

$$\dot{I} = 0, \quad \dot{\phi} = \omega(I), \text{ where } \omega(I) = \frac{\partial H}{\partial I},$$

and the motion on M is quasiperiodic since $H = F_1 = H(I)$ and Equations (A 26.1), in action-angle coordinates, are Hamiltonian equations[2] with corresponding Hamiltonian function $H(I)$.

Proof:

NOTATIONS A 26.4

We use the following notations. Let $x = (p, q)$ be a point of the phase space R^{2n}; we shall denote by grad F the vector gradient $F_{x_1}, \ldots, F_{x_{2n}}$ of a function $F(x)$. The Hamiltonian equations (A 26.1) then take the form:

$$\text{(A 26.5)} \qquad \dot{x} = I \operatorname{grad} H, \qquad I = \begin{pmatrix} O & -E \\ E & -O \end{pmatrix},$$

where E is the unit matrix of order n. We introduce in R^{2n} the skew-scalar product of two vectors x, y:

$$[x, y] = (Ix, y) = -[y, x],$$

where (,) is the usual scalar product. As can be easily verified, $[x, y]$ expresses the sum of the areas of the projections of the parallelogram with sides x, y onto the coordinate planes $p_i q_i$ $(i = 1, \ldots, n)$.

Linear transformations S, which preserve the skew-scalar product

$$[Sx, Sy] = [x, y] \text{ for all } x, y,$$

[2] See Appendix 32.

are called symplectic. For instance, the transformation with matrix I is symplectic. The skew-scalar product of the gradients $[\mathrm{grad}\, F, \mathrm{grad}\, G]$ is called the Poisson bracket (F, G) of the functions F, G. Obviously, F is a first integral of the system (A 26.5) if and only if its Poisson bracket (F, H) with the Hamiltonian vanishes identically. If the Poisson bracket of two functions vanishes identically, the functions are said to be *in involution*.

THE CONSTRUCTION. A 26.6

Consider the n vector fields: $\xi_i = I\, \mathrm{grad}\, F_i$, $(i = 1, \ldots, n)$. On account of the nondegeneracy of I and the linear independence of the $\mathrm{grad}\, F_i$'s, the vectors ξ_i are *linearly independent* at each point of M.

Let us consider the system (A 26.5) with Hamiltonian F_i. Since $(F_i, F_j) = 0$, all the functions F_j are first integrals, and every orbit lies wholly on M. *Therefore the velocity field* $\xi_i = I\, \mathrm{grad}\, F_i$ *is tangent to* M.

Finally, *the fields* ξ_j and ξ_i *commute*, for their Lie bracket is nothing but[3] the velocity field of the system (A 26.5) with Hamiltonian $(F_i, F_j) = 0$.

Thus, M is a connected, compact orbit of the group $\mathbf{R}^n$ acting smoothly and transitively; therefore we proved that $M = T^n$. Besides, M being specified by the equations $F_i = f_i = \text{constant}$, the fields $\mathrm{grad}\, F_i$ define a structure of direct product in the neighborhood of M.

Now, let us choose the torus M_f: $F = f$ in the neighborhood of M and consider the n integrals

$$I_i(f) = \frac{1}{2\pi} \oint_{\gamma_i(f)} p\, dq \tag{A 26.7}$$

over the basic cycles $\gamma_i(f)$ of the torus M_f. Since the $I_i(f)$ are functionally independent, the equation can be solved in the neighborhood of M and yields the torus $M(I) = M_{f(I)}$, which corresponds to a given I. Let us set:

---

[3] The Lie bracket of the Hamiltonian vector fields $I\, \mathrm{grad}\, F$ and $I\, \mathrm{grad}\, G$ is an Hamiltonian vector field with Hamiltonian function $-(F, G)$. We shift the proof to Appendix 32.

(A 26.8) $$S(I, q) = \int_{q_0}^{q} p\,dq$$

where the path of integration lies on $M(I)$ (therefore, $p = p(I, q)$). The many-valued function S is the generating function (see Appendix 32) of the canonical transformation $I, \phi \to p, q$, which defines action-angle coordinates:

(A 26.9) $$p = \frac{\partial S}{\partial q}, \quad \phi = \frac{\partial S}{\partial I} .$$

LEMMA A 26.10

The one-form $p\,dq$ of $M(I)$ is closed.

Proof:

It is sufficient to prove that the integral of $p\,dq$ along infinitely small parallelograms lying in $M(I)$ vanishes. If D is a parallelogram with sides ξ, η, then $\oint_D p\,dq$ (i.e., the sum of the areas of the projections of D onto the coordinate planes $p_i q_i$, $i = 1, \dots, n$) is the skew-product $[\xi, \eta]$ of ξ and η. Suppose now that ξ and η touch $M(I)$ at a certain point. In accordance with (A 26.6) any vector tangential to $M(I)$ is a linear combination of the n vectors I grad F_i. But these vectors are skew-orthogonal since, in accordance with (A 26.2),

$$[\text{grad } F_i, \text{grad } F_j] = 0 ,$$

and thus, since I is symplectic,

$$[I \text{ grad } F_i, I \text{ grad } F_j] = 0 .$$

Therefore $[\xi, \eta] = 0$, as required. The integral (A 26.8) can therefore be regarded as a many-valued function S and Equations (A 26.9) define, locally, a canonical transformation $I, \phi \to p, q$.

ACTION-ANGLE VARIABLES A 26.11

In fact formulas (A 26.9) define a global canonical mapping in which p and q have period 2π with respect to ϕ. To prove it, we observe that, for every I, the differential of $S(I, q)$ is a global one-form on $M(I)$. Therefore $d\phi$, as defined by (A 26.9), is also a global one-form.

Let us compute the periods of the one-forms $d\phi_i$ over the basic cycles of the torus M_f. According to (A 26.7) we have:

$$\oint_{\gamma_j} d\phi_i = \oint_{\gamma_j} d\left(\frac{\partial S}{\partial I_i}\right) = \frac{d}{dI_i} \oint_{\gamma_j} dS = \frac{d}{dI_i}(2\pi I_j) = 2\pi\delta_{ij}$$

Therefore the variables ϕ_i are angular coordinates on the torus $M(I)$ and our theorem is proved.

APPENDIX 27

SYMPLECTIC LINEAR MAPPINGS OF PLANE

(See Section 20, Chapter 4)

Let A be a symplectic linear mapping of the plane (p, q). The mapping A preserves the area-element $dp \wedge dq$, therefore Det A = 1. Consequently, the product of the proper values λ_1 and λ_2 of A is equal to 1. Besides, λ_1 and λ_2 are roots of the real polynomial $\text{Det}(A - \lambda E)$. Therefore, either λ_1 and λ_2 are both real, or they are complex conjugate: $\lambda_2 = \bar{\lambda}_1$.

In the *first case*, we have:

(A 27.1) $$|\lambda_2| < 1 < |\lambda_1| .$$

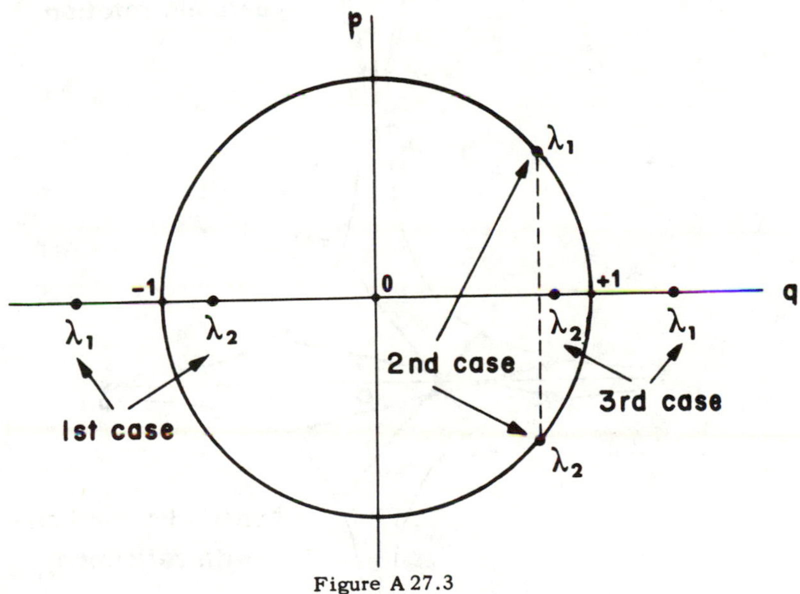

Figure A 27.3

In the *second* case, we have:

$$1 = \lambda_1 \lambda_2 = \lambda_1 \bar{\lambda}_1 = |\lambda_1|^2 = |\lambda_2|^2, \qquad \lambda_1 \neq \lambda_2 , \tag{A 27.2}$$

and the roots belong to the unit circle (see Figure A 27.3). The *third* and last possible *proper value configuration is*:

$$\lambda_1 = \lambda_2 = \pm 1 .$$

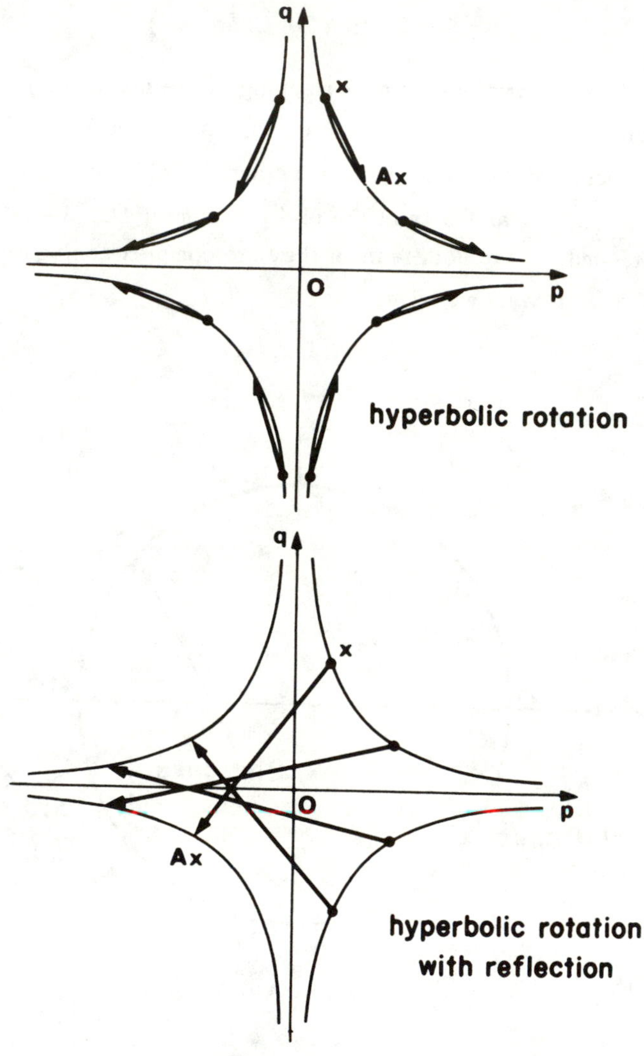

Figure A 27.5

EXAMPLE A 27.4

The hyperbolic rotation:

$$p, q \to 2p, \tfrac{1}{2} q ,$$

or the *hyperbolic rotation with reflection:*

$$p, q \to -2p, -\tfrac{1}{2} q \quad \text{(see Figure A 27.5).}$$

In both cases the orbit $T^n x$ of $x = (p, q)$ belongs to the hyperbola $pq =$ constant. Of course, the fixed point O is unstable. From classical theorems of linear algebra it follows that every mapping $\mathbf{A}$ of the first type ($\lambda_1 \neq \lambda_2$; $\lambda_1, \lambda_2 \in \mathbf{R}$) is an hyperbolic rotation, possibly with reflection. This means that, up to a suitable change of variables, $\mathbf{A}$ may be written under the form:

$$P, Q \to \lambda P, \frac{1}{\lambda} Q .$$

EXAMPLE A 27.6

A *rotation through an angle* α belongs to the second class ($\lambda_1 = e^{-i\alpha}$, $\lambda_2 = e^{i\alpha}$):

$$p, q \to p \cos \alpha - q \sin \alpha, \; p \sin \alpha + q \cos \alpha .$$

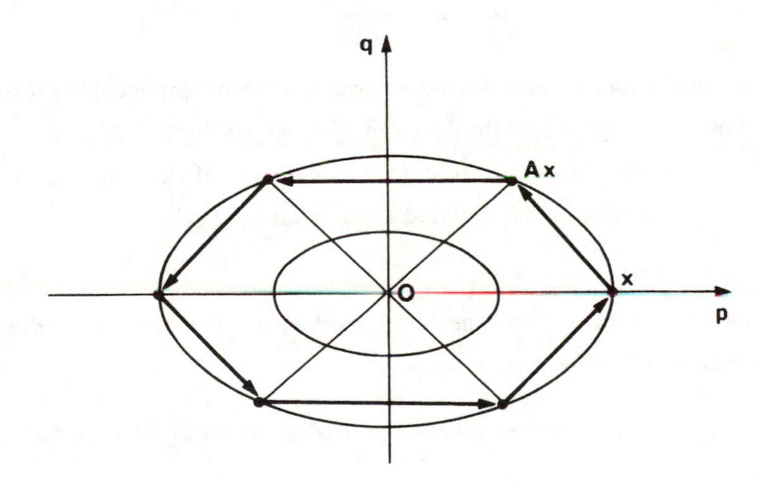

Figure A 27.7

This rotation transforms into an "elliptic rotation" (see Figure A 27.7) under a linear change of variables. In this case the orbit $T^n x$ of $x = (p, q)$ belongs to an ellipse centered at O. The fixed point O is obviously stable. Classical theorems of linear algebra show that every mapping A of the second type $(|\lambda_1| = |\lambda_2| = 1, \lambda_1 \neq \lambda_2)$ is an elliptic rotation.

In the first case (A 27.1), the fixed point O is called an *hyperbolic point* and one says that A is *hyperbolic at O*. In the second case (A 27.2), the fixed point O is called an *elliptic point* and one says that A is *elliptic at O*. Finally, the third case $(\lambda^2 = 1)$ is called the *parabolic case*.

REMARK A 27.8

Every canonical mapping A′, close enough to an elliptic mapping A, is elliptic. In fact, the roots λ_1 and λ_2 depend continuously on A and are restricted to lie either on the real axis or on the unit circle (see Figure A 27.3). Therefore, these roots cannot leave the unit circle, except at points $\lambda = \pm 1$ which correspond to the parabolic case. Finally, we define the *topological index* of a vector field at a fixed point. Let us consider a vector field $\xi(x)$ of the plane p, q, with an isolated fixed point $\xi(0) = 0$. This defines a mapping of the unit circle $x^2 = p^2 + q^2 = 1$ onto itself:

$$B(\varepsilon): x \to \frac{\xi(\varepsilon \cdot x)}{\|\xi(\varepsilon \cdot x)\|} .$$

If ε is small enough, then the topological degree of this mapping does not depend on ε and is called the index of ξ at O, or the index of O.

Now, consider the vector field $\xi(x) = Ax - x$. If the mapping A is nonparabolic, then O is an isolated fixed point of $\xi(x)$.

THEOREM A 27.9

An elliptic point, or an hyperbolic point with reflection, has index $+1$. *An hyperbolic point has index* -1.

Proof consists in a mere inspection of Figures (A 27.5) and (A 27.7).

APPENDIX 28

STABILITY OF THE FIXED POINTS

(See Section 20, Chapter 4)

Consider an analytical canonical mapping A of the plane $p, q,$ with fixed point $O = (0, 0)$. Assume that O is elliptic, that is, that the differential of A at zero has proper values $\lambda_1 = e^{-i\alpha}$, $\lambda_2 = e^{i\alpha}$. It has been known since G. D. Birkhoff's time[1] that, if $\alpha/2\pi$ is irrational, to every $s > 0$ corresponds a canonical mapping $B = B(s)$ of a neighborhood of O:

$$B: \; p, q \to P, Q, \qquad B(O) = O,$$

which reduces A to a "normal form":

$$A' = BAB^{-1}: \; P, Q \to P', Q',$$

that is as follows. Let I, ϕ be the canonical polar coordinates:

$$2I = P^2 + Q^2, \qquad \phi = \operatorname{arc\,tg}(P/Q)$$
$$2I' = P'^2 + Q'^2, \qquad \phi' = \operatorname{arc\,tg}(P'/Q'),$$

then:

$$I' - I = 0(I^{s+1}) \tag{A 28.1}$$
$$\phi' - \phi = \alpha + \alpha_1 I + \alpha_2 I^2 + \cdots + \alpha_s I^s + 0(I^{s+1}) .$$

The coefficients $\alpha, \alpha_1, \ldots$ do not depend on the mapping $B(s)$ by which A is reduced to the form A$'$. If $\alpha \neq 2\pi m/n$ and if there is a nonvanishing coefficient $\alpha_1, \alpha_2, \ldots,$ Birkhoff says that A is of "generic elliptic type."

[1] *Dynamical Systems,* Chapter 3.

THEOREM A 28.2 (See Arnold [7]).

The fixed point of a generic elliptic mapping is stable.

The proof consists in applying the construction of Theorem (21.11) of Chapter 4 (see Appendix 34) to the mapping (A 28.1): for $I \ll 1$, $0(I^{s+1})$ is regarded as a perturbation of the mapping

$$I' = I, \qquad \phi' = \phi + a + a_1 I + \cdots + a_s I^s .$$

Similar theorems are obtained concerning the stability of equilibrium positions and elliptic periodic solutions of Hamiltonian systems with two degrees of freedom (see Arnold [7]). J. Moser [1] obtained the strongest result in that way:

MOSER'S THEOREM A 28.3

The fixed point of an elliptic canonical mapping A *of the plane is stable provided that:*

(1) $a \neq 2\pi \frac{m}{3}, \quad 2\pi \frac{m}{4}$;

(2) $a_1 \neq 0$;

(3) A *is* C^{333}*-differentiable.* (As it is pointed out in a recent paper of Moser [6], this number of derivatives can be fairly well reduced.) A complete proof will be found in J. Moser [1].

REMARK A 28.4

If $a = 2\pi m/3$, then the fixed point can be unstable, as shown by Levi-Civita [1].

APPENDIX 29

PARAMETRIC RESONANCES

(See Section 20, Chapter 4)

The analysis of the stability of the fixed point $(0,0)$ of a linear mapping of the plane is due to Poincaré and Lyapounov. Only in recent times (1950), were these results extended by M. G. Krein [1], [2] to systems with many degrees of freedom. Krein's investigations have been enlarged by Jacoubovich [1], Gelfand and Lidskii [4], and so on. J. Moser [3] published a report of Krein's theorem.

Let A be a linear symplectic mapping[1] of the canonical space R^{2n}. We say that A is *stable if the sequence* A^n *is bounded.* We say that A is *parametrically stable* if every symplectic mapping, close to A, is stable. We proved in Appendix 27 (and used it in Section 20, Chapter 4) that every elliptic mapping of R^2 is parametrically stable. M. G. Krein displayed all the parametrically stable mappings of R^{2n}.

LEMMA A 29.1 (Poincaré-Lyapounov)

Suppose A *is a symplectic mapping and* λ *is a proper value of* A. *Then* $1/\lambda$, $\bar{\lambda}$, *and* $1/\bar{\lambda}$ *are proper values of* A.

---

[1] A preserves the skew-scalar product $[\xi,\eta] = (I\xi,\eta)$, where $(\ ,\ )$ is the inner product and $I = \begin{pmatrix} 0 & -E \\ E & -0 \end{pmatrix}$, E = unit matrix of order n. Therefore, we have:

$$[\mathrm{A}\xi, \mathrm{A}\eta] = [\xi,\eta] \text{ and } \mathrm{A}'I\mathrm{A} = I.$$

Proof:

It is sufficient to prove that the characteristic polynomial of $\mathbf{A}$ is real and reciprocal. In fact, we have:

$$\begin{aligned} p(\lambda) &= \mathrm{Det}(\mathbf{A}-\lambda E) = \mathrm{Det}(-I\mathbf{A}'^{-1}I + \lambda I^2) \\ &= \mathrm{Det}(-\mathbf{A}'^{-1} + \lambda E) = \mathrm{Det}(-\mathbf{A}^{-1} + \lambda E) \\ &= \mathrm{Det}(-E + \lambda\mathbf{A}) = \lambda^{2n}\cdot\mathrm{Det}(\mathbf{A}-\lambda^{-1}E) = \lambda^{2n}\cdot p(\lambda^{-1}) \,. \end{aligned}$$

From this lemma the following corollary is readily deduced:

COROLLARY A 29.2

The proper values of $\mathbf{A}$ *divide into couples and "quadruples." Couples are formed by* λ *and* λ^{-1}, λ *belonging to the real axis or the unit circle:* $|\lambda| = 1$. *Quadruples are formed by* $\lambda, \bar{\lambda}, \lambda^{-1}$, *and* $\bar{\lambda}^{-1}$ (see Figure A 29.3).

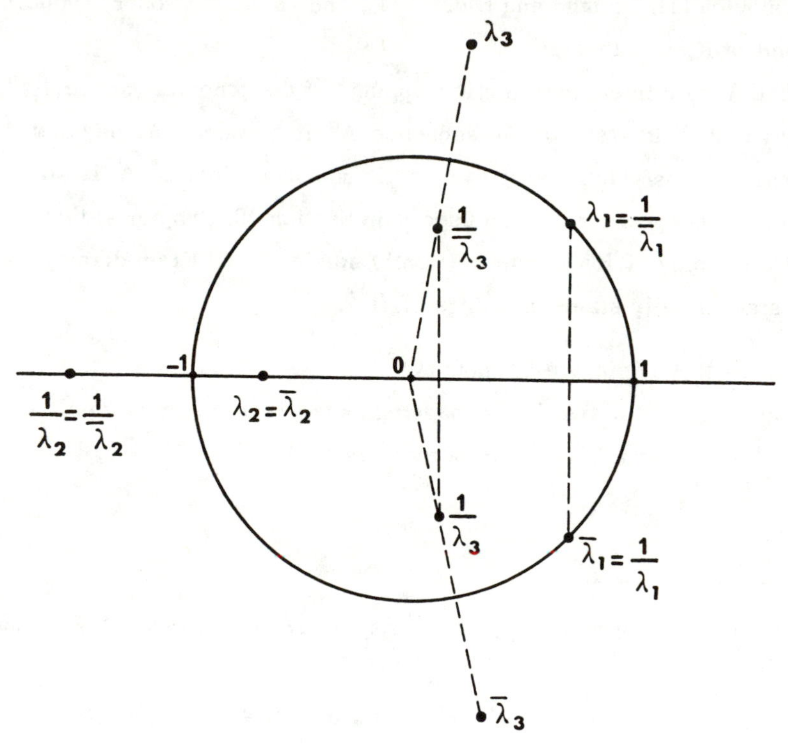

Figure A 29.3

COROLLARY A 29.4

If the proper values are simple and lie on the unit circle $|\lambda| = 1$, *then* A *is parametrically stable*, because if all the proper values are simple and lie on $|\lambda| = 1$, then:

(1) A is stable (for obvious reasons of normal form);

(2) all the proper values of a symplectic mapping A′, close enough to A, lie on $|\lambda| = 1$. In fact, assume the contrary, then A′ would have two proper values λ and $\bar{\lambda}^{-1}$ close to a unique isolated proper value of A. (see Figure A 29.3).

Let us now assume definitively that ± 1 are not proper values of A. Krein classified the proper values belonging to the unit circle $|\lambda| = 1$: they split into positive and negative proper values. First assume that all the proper values are simple; we prove the following lemma:

LEMMA A 29.5

Let ξ_1 *and* ξ_2 *be the proper vectors with corresponding proper values* λ_1 *and* λ_2 . *Then, either* $\lambda_1\lambda_2 = 1$, *or* $[\xi_1, \xi_2] = 0$.

Proof:

Since $A\xi_1 = \lambda_1\xi_1$ and $A\xi_2 = \lambda_2\xi_2$, we have:

$$[A\xi_1, A\xi_2] = \lambda_1\lambda_2 \, [\xi_1, \xi_2] = [\xi_1, \xi_2] \, .$$

Consequently, we have: $(\lambda_1\lambda_2 - 1)[\xi_1, \xi_2] = 0$.

(Q. E. D.)

COROLLARY A 29.6

Let σ *be a plane, invariant under* A *and corresponding to conjugate proper values* $\lambda_1, \lambda_2, |\lambda_1| = |\lambda_2| = 1$. *Then:*

(1) σ *is skew-orthogonal to every proper vector* ξ_3 *corresponding to another proper value* λ_3;

(2) *the skew-product* $[\xi, \eta]$ *of noncolinear vectors* ξ *and* η *of* σ *is nonvanishing*.

Assertion (1) is a direct consequence of $\lambda_1\lambda_3 \neq 1$, $\lambda_2\lambda_3 \neq 1$: in view of Lemma (A 29.5) we have $[\xi_1, \xi_3] = [\xi_2, \xi_3] = 0$. Suppose $[\xi_1, \xi_2] = 0$. We have also $[\xi_1, \xi_1] = 0$, and assertion (1) implies $[\xi_1, \xi_3] = 0$ for every ξ_3. Therefore $[\xi_1, \eta] = 0$ for every η, which is impossible. Consequently $[\xi_1, \xi_2] \neq 0$ and assertion (2) holds good.

DEFINITION A 29.7

A proper value λ, such that $|\lambda| = 1$, $\lambda^2 \neq 1$, is called a positive (resp. negative) proper value of A *if: $[A\xi, \xi] > 0$ (resp. < 0) for every ξ of the real invariant plane σ corresponding to the proper values λ and $\bar{\lambda}$.*

This definition is correct. Indeed, the vectors $A\xi$ and ξ of σ are noncolinear because $\lambda^2 \neq 0$. Therefore, in view of Corollary (A 29.6), $[A\xi, \xi] \neq 0$ on σ. Consequently $[A\xi, \xi]$ has constant sign for every $\xi \in \sigma$.

REMARK A 29.8

The sign of a proper value has a simple geometrical meaning. The plane σ admits a canonical orientation, for $[\xi, \eta] \neq 0$ if ξ is nonparallel to η. Therefore, one may speak of positive (or negative) rotations. The restriction of A to σ is an elliptic rotation through an angle a, $0 < |a| < \pi$. *The proper value λ is positive (resp. negative) if* A *rotates σ through a positive (resp. negative) angle.*

Krein's main result is: *collision of two proper values with identical signs on the unit circle $|\lambda| = 1$ does not provoke instability.* In contrast, two proper values with opposite signs can leave the unit circle after they have collided, so forming a "quadruple" with their conjugates (see Figure A 29.3).

To be precise, let $A(t)$ be a symplectic mapping which depends continuously on a parameter t, and the proper values of which are different from ± 1 if $|t| < \tau$. Suppose that, for $t < 0$, all the proper values λ_k of A are simple and lie on the unit circle, while certain of these proper values collide for $t = 0$.

THEOREM A 29.9

If all the proper values that collide have identical sign, then they re-

main on the unit circle after the collision and the mapping A *remains stable for* $t < \varepsilon$, $\varepsilon > 0$.

We shall prove the theorem in the simplest case in which all the proper values λ, $\mathfrak{J}\lambda > 0$ collide. The general case can be reduced to this case by selecting a canonical subspace $R^{2l}(t)$ corresponding to the l colliding proper values and their conjugates. To fix the ideas, suppose that the proper values λ_k are positive:

$$[A\xi, \xi] > 0 \text{ for } \xi \in \sigma_k ,$$

where σ_k is the plane generated by ξ_k, $\bar{\xi}_k$ $(A\xi_k = \lambda_k \xi_k)$.

Proof of the Theorem. A 29.10

Consider the quadratic form $[A\xi, \xi]$; its polar bilinear form is nondegenerate. We have, indeed:

$$[A\xi, \eta] + [A\eta, \xi] = [A\xi, \eta] - [A^{-1}\xi, \eta] = [(A - A^{-1})\xi, \eta] .$$

Suppose $[(A - A^{-1})\xi, \eta] = 0$ for every η, then $(A - A^{-1})\xi = 0$ and $(A^2 - E)A\xi = 0$. Thus, 1 would be a propervalue of A^2, which contradicts the condition of Theorem (A 29.9) $(\lambda \neq \pm 1)$. Therefore $[A(t)\xi, \xi]$ is *nondegenerate* for $|t| < \tau$ and, in particular, for $t = 0$. On the other hand, *this form is positive definite* for $t < 0$. In fact, every vector η is equal to the sum of its projections η_k into the invariant planes σ_k corresponding to the proper values λ_k, $\bar{\lambda}_k$. According to Lemma (A 29. 6) these planes σ_k are skew-orthogonal, therefore:

$$[A\eta, \eta] = \sum_{k,l} [A\eta_k, \eta_l] = \sum_k [A\eta_k, \eta_k] ,$$

because

$$[A\eta_k, \eta_l] = 0 \text{ if } k \neq l \quad (A\eta_k \in \sigma_k, \ \eta_l \in \sigma_l) .$$

But $[A\eta_k, \eta_k] > 0$ for λ_k is a positive proper value, hence:

$$[A\eta, \eta] > 0 .$$

So, *the form* $[A(t)\xi, \xi]$ *is positive definite for* $t < 0$ *and nondegenerate* for $t = 0$. Therefore, this form is positive definite for $t = 0$ and, consequently, for $t < \varepsilon$, $\varepsilon > 0$. But $[A\,A^n\xi, A^n\xi] = [A\xi, \xi]$ for A^n is symplectic. Thus the orbit $A^n\xi$ belongs to the ellipsoid $[A\xi, \xi]$ = constant, that is $A(t)$ is stable for $t < \varepsilon$. (Q. E. D.)

REMARK A 29.11

The above argument proves the criterion of parametric stability: *The symplectic mapping* A *is parametrically stable if and only if all the proper values* λ_k *lie on the unit circle* $|\lambda_k| = 1$, $\lambda_k^2 \neq 1$. *Besides, the quadratic form* $[A\xi, \xi]$ *is definite on every invariant subspace corresponding to the multiple proper values* λ_k, $\bar{\lambda}_k$.

APPENDIX 30

THE AVERAGING METHOD FOR PERIODIC SYSTEMS

(See Section 22, Chapter 4)

Let $\Omega = B^l \times S^1$ be the phase space, where $B^l = \{I = (I_1, \ldots, I_l)\}$ is an open bounded subset of R^l and $S^1 = \{\phi \,(\mathrm{mod}\ 2\pi)\}$ is a circle. We consider ϕ-periodic smooth functions $\omega(I)$, $F(I,\phi)$, $f(I,\phi)$:

$$F:\ \Omega \to R^l, \qquad f:\ \Omega \to R^l, \qquad \omega:\ B^l \to R^l,$$

and finally, $\varepsilon \ll 1$ denotes a small parameter.

THEOREM A 30.1

We consider the systems defined in Ω*:*

$$\text{(A 30.2)} \qquad \begin{cases} \dot{\phi} = \omega(I) + \varepsilon f(I,\phi) \\ \dot{I} = \varepsilon \cdot F(I,\phi) \ , \end{cases}$$

and

$$\text{(A 30.3)} \qquad \dot{J} = \varepsilon \cdot \bar{F}(J), \quad \text{where} \quad \bar{F}(J) = \frac{1}{2\pi} \int_0^{2\pi} F(J,\phi)\, d\phi \ .$$

If $\omega(I) \neq 0$ *in* Ω, *then the solutions* $I(t)$ *and* $J(t)$ *of* (A 30.2) *and* (A 30.3), *with equal initial data* $I(0) = J(0)$, *satisfy*[1]*:*

$$|I(t) - J(t)| < C \cdot \varepsilon \quad \textit{for every } t,\ 0 \leq t \leq 1/\varepsilon \ ,$$

where C *is a constant which does not depend on* ε .

---

[1] We suppose that $J(t) \in B^l$ for every t, $0 \leq t \leq 1/\varepsilon$.

Proof:

Let us improve (A 30.2) by using a new variable:

(A 30.4) $$P = P(I, \phi) = I + \varepsilon g(I, \phi), \qquad g: \Omega \to R^1 ,$$

From (A 30.2) and (A 30.4) follows:

(A 30.5) $$\dot{P} = \varepsilon F(P, \phi) + \varepsilon \frac{\partial g(P, \phi)}{\partial \phi} \omega(P) + O(\varepsilon^2) .$$

In order to cancel the terms of order ε, we set:

(A 30.6) $$g(I, \phi) = \int_0^{\phi} \frac{\bar{F}(P) - F(P, \phi)}{\omega(P)} d\phi ;$$

this expression is well-defined since

$$\omega(P) \neq 0, \quad \int_0^{2\pi} (\bar{F} - F) d\phi = 0,$$

and therefore $g(\phi + 2\pi) = g(\phi)$. Now, our system (A 30.5) may be written:

(A 30.7) $$\dot{P} = \varepsilon \bar{F}(P) + O(\varepsilon^2) .$$

Let $P(t)$ be the solution of (A 30.7) with initial data $P(0) = J(0) = I(0)$:

(A 30.8) $$P(t) \equiv P(I(t), \phi(t)) .$$

Obviously, (A 30.7) implies:

(A 30.9) $$|P(t) - J(t)| < C_1 \cdot \varepsilon \quad \text{for every } t, \ 0 \leq t \leq 1/\varepsilon .$$

Finally, from (A 30.4), (A 30.6) and (A 30.8) follows:

(A 30.10) $$|P(t) - I(t)| < C_2 \cdot \varepsilon \quad \text{for every } t .$$

Inequalities (A 30.9) and (A 30.10) conclude the proof. They prove also that the motion decomposes into the averaged motion and fast small oscillations (see Figure A 30.11).

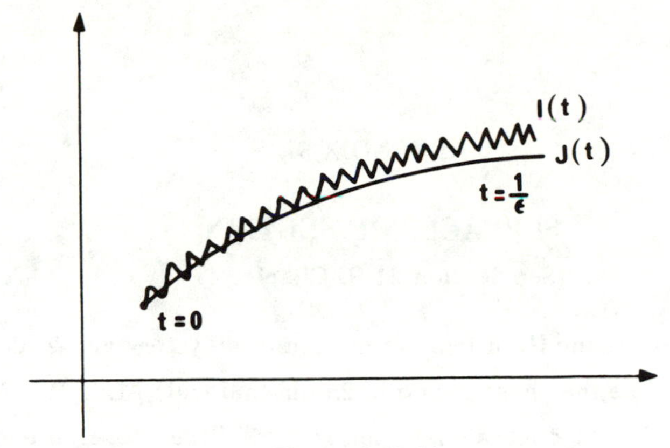

Figure A 30.11

APPENDIX 31

SURFACES OF SECTION

(See Section 21.9, Chapter 4)

Let $H(p, q)$ be the Hamiltonian function of a system with n degrees of freedom (therefore the phase space is $2n$-dimensional). Let Σ: $H = h$, $q_1 = 0$ be a $(2n-2)$-dimensional submanifold of the "level of energy" $H = h$. If, in a certain domain Σ_0 of Σ, $P = (p_2, \ldots, p_n)$, $Q = (q_2, \ldots, q_n)$ form a local chart and $\dot{q}_1 \neq 0$, Σ is called a surface of section (see Figure A 31.1). Assume that an orbit of the Hamiltonian system, through a

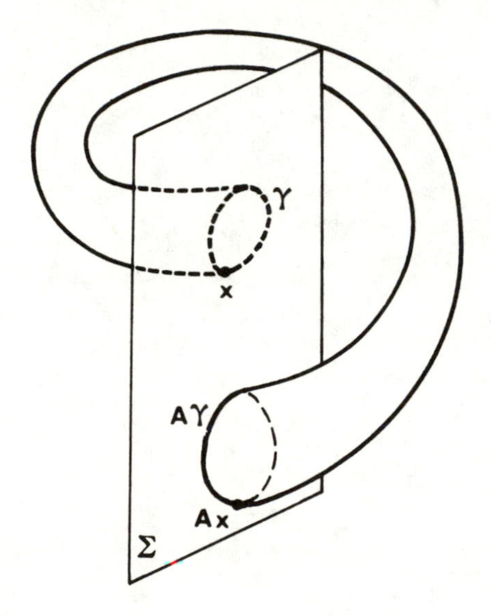

Figure A 31.1

point x of Σ_0, returns to Σ_0. Then, in view of $\dot{q}_1 \neq 0$, the orbit through a point x' on Σ_0 sufficiently close to x, will, as t increases, return to

Σ_0 and will cut Σ_0 in a uniquely determined point Ax'. In this manner, we define a mapping A:

$$\Sigma_1 \to \Sigma_0, \quad \Sigma_1 \subset \Sigma_0 \subset \Sigma .$$

THEOREM [1] A 31.2

The mapping A *is canonical, that is for every closed curve* γ *of* Σ_1, *we have:*

(A 31.3)
$$\oint_{\gamma} PdQ = \oint_{A\gamma} PdQ ,$$

where $PdQ = p_2 dq_2 + \cdots + p_n dq_n$.

Proof:

Consider the orbits emanating from γ in the $(2n+1)$-dimensional space $\{(p, q, t)\}$. The curves γ and $A\gamma$ of the space $\{(p, q)\}$ are the projections of two closed curves γ' and $A\gamma'$ of $\{(p, q, t)\}$ which are formed respectively,

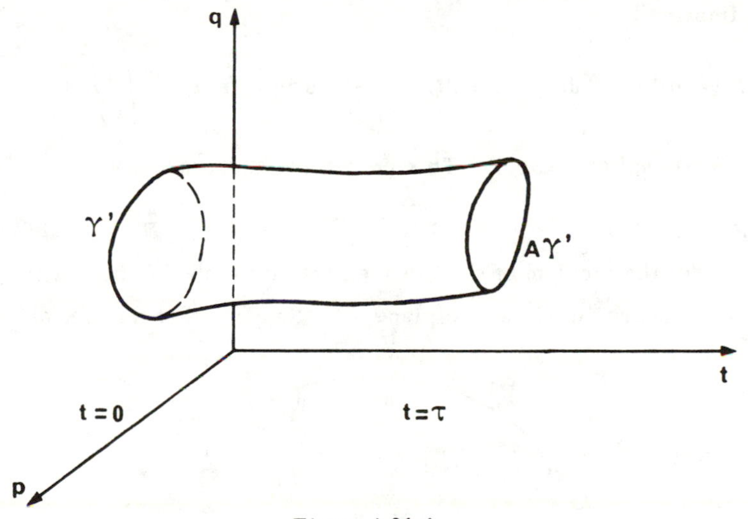

Figure A 31.4

by the initial points ($t = 0$) and the end points of the above orbits (see Figure A 31.4). Therefore, we have by the Poincaré-Cartan theorem:

---

[1] A proof of this well-known theorem has apparently never been published.

(A 31.5) $$\oint_{\gamma'} p\,dq - H\,dt = \oint_{A\gamma'} p\,dq - H\,dt ,$$

where $p\,dq = p_1\,dq_1 + \cdots + p_n\,dq_n$. But H = constant along γ' and $A\gamma'$; then:

$$\oint_{\gamma'} H\,dt = \oint_{A\gamma'} H\,dt = 0 .$$

Besides, we have:

$$\oint_{\gamma'} p\,dq = \oint_{\gamma} p\,dq \quad \text{and} \quad \oint_{A\gamma'} p\,dq = \oint_{A\gamma} p\,dq .$$

In view of q_1 = constant on Σ, we also have:

$$\oint_{\gamma'} p_1\,dq_1 = \oint_{A\gamma'} p_1\,dq_1 = 0 .$$

Thus, finally,

$$\oint_{\gamma'} p\,dq - H\,dt = \oint_{\gamma} P\,dQ, \qquad \oint_{A\gamma'} p\,dq - H\,dt = \oint_{A\gamma} P\,dQ ,$$

and (A 31.5) implies (A 31.3). This proves the theorem.

EXAMPLE A 31.6

Consider the problem of the "convex billiard table" (Birkhoff). Let Γ be a closed convex curve of the plane E^2. Suppose that a material point

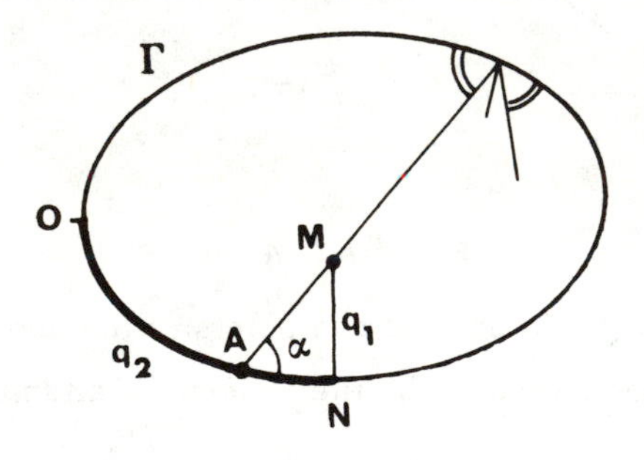

Figure A 31.7

M moves inside Γ and collides with Γ according to the law "the angle of incidence is equal to the angle of reflection" (see Figure A 31.7). The states of M, immediately before and immediately after a reflection, are determined by the angle of incidence α, $0 \leq \alpha \leq 2\pi$, and the point of incidence. The point of incidence A is defined by the algebraic length q_2 of the arc OA of Γ (O is an arbitrary origin). In other words, the set of the states of M, immediately before and immediately after a reflection, form a torus $T^2 = \{\alpha(\text{mod } 2\pi),\ q_2 \ (\text{mod } L)\}$ in the phase space (L is the length of Γ). We obtain naturally a mapping $\mathbf{A}$ of a subset of T^2 into another one: the state which immediately follows a reflection is transformed into the state immediately preceding the next reflection.

THEOREM A 31.8 (G. D. Birkhoff)

$$I = \sin\alpha \cdot dq_2 \wedge d\alpha \text{ is invariant under the mapping } \mathbf{A}.$$

Proof:

Between two reflections, the motion of M is determined by Hamiltonian equations in the corresponding four-dimensional phase-space. In the neighborhood of the above torus T^2, let us select special coordinates. Our point M is well-defined by coordinates (q_1, q_2), where $q_1 = MN$ is the distance from M to Γ and q_2 is the algebraic length of the arc ON.

Coordinates q_1 and q_2 (mod L) are clearly Lagrangian coordinates in a neighborhood of Γ. Let p_1 and p_2 be the corresponding momenta (the mass of M is supposed to be 1). On Γ, p_1 and p_2 coincide obviously with the components of the velocity-vector $\mathbf{v}$:

$$p_1 = |v| \cdot \sin\alpha, \quad p_2 = |v| \cdot \cos\alpha .$$

The Hamiltonian function H is the kinetic energy:

$$H = \frac{v^2}{2} .$$

In the four-dimensional space $\{(p_1, p_2, q_1, q_2)\}$, consider the surface Σ whose equation is:

$$H = \tfrac{1}{2}, \quad q_1 = 0 \ (\text{i.e., } |v| = 1,\ M \in \Gamma):$$

From one reflection to the next, the motion defines a mapping $A: \Sigma \to \Sigma$. The coordinates p_2, q_2 are local coordinates of Σ $(\alpha \neq 0)$ which is a surface of section. In view of Theorem (A 31.2), the mapping A is canonical and therefore preserves the two-form:

$$dp_2 \wedge dq_2 = \sin \alpha \cdot dq_2 \wedge d\alpha .$$

(Q. E. D.)

An elementary proof of this theorem, due to G. D. Birkhoff [1], requires extensive computations.

APPENDIX 32

THE GENERATING FUNCTIONS OF CANONICAL MAPPINGS

(See Section 21, Chapter 4)

The following results are due to Hamilton and Jacobi.

§A. Finite Canonical Mappings

Let $x = (p, q)$, $(p = (p_1, \ldots, p_n),\ q = (q_1, \ldots, q_n))$, be a point of the canonical space R^{2n}. The differentiable mapping:

$$\mathrm{A}: \ x \to X = (P(p, q), Q(p, q)),\ (P = (P_1, \ldots, P_n),\ Q = (Q_1, \ldots, Q_n)) \,.$$

is called *canonical* if A preserves the Poincaré integral-invariant:

$$\oint_{\gamma} p\,\mathrm{d}q = \oint_{\mathrm{A}\gamma} p\,\mathrm{d}q \,, \tag{A 32.1}$$

for any closed curve γ. Let σ be an arbitrary two-chain. Relation (A 32.1) implies that A preserves the sum of the areas of the projections of σ into the coordinate planes p_i, q_i:

$$I(\sigma) = \iint_{\sigma} \mathrm{d}p \wedge \mathrm{d}q = \iint_{\mathrm{A}\sigma} \mathrm{d}p \wedge \mathrm{d}q = I(\mathrm{A}\sigma) \,. \tag{A 32.2}$$

In other words, the two forms $\mathrm{d}p \wedge \mathrm{d}q$ and $\mathrm{d}P \wedge \mathrm{d}Q$ coincide:

$$\mathrm{d}p \wedge \mathrm{d}q = \mathrm{d}P \wedge \mathrm{d}Q, \ \text{where } P = P(p, q),\ Q = Q(p, q) \,. \tag{A 32.3}$$

If the domain of A is simply connected, then conditions (A 32.1) and (A 32.2) are equivalent. Relation (A 32.3) shows that:

$$p\,dq + Q\,dP, \text{ where } P = P(p, q), \quad Q = Q(p, q),$$

is a closed form of R^{2n} (since $dp \wedge dq + dQ \wedge dP = 0$). Therefore:

$$A(x) = \int_{x_0}^{x} p\,dq + Q\,dP, \text{ where } P = P(p, q), \quad Q = Q(p, q),$$

defines, locally, a function on R^{2n}. Suppose that $q_1, \ldots, q_n;\ P_1, \ldots, P_n$, form a local chart in some neighborhood of the point x, that is:

$$\mathrm{Det}\left(\frac{\partial P}{\partial p}\right) \neq 0 .$$

Then, $A(x)$ can be regarded as a function of P, q, defined in the neighborhood of the point P, q:

$$\text{(A 32.4)} \quad A(P, q) = \int^{(P,q)} p\,dq + Q\,dP, \text{ where } p = p(P, q), \quad Q = Q(P, q).$$

DEFINITION A 32.5

The function $A(P, q)$ is called the generating function of the canonical mapping **A**.

Of course, A is only defined locally and up to a constant. From (A 32.4) follows:

$$\text{(A 32.6)} \qquad \frac{\partial A}{\partial P} = Q, \quad \frac{\partial A}{\partial q} = p .$$

LEMMA A 32.7

Let $A(P, q)$ be a function such that:

$$\mathrm{Det}\left(\frac{\partial^2 A}{\partial P\,\partial q}\right) \neq 0$$

in the neighborhood of a point (P, q). Then, Equations (A 32.6) can be solved locally with respect to P and Q:

$$P = P(p, q), \quad Q = Q(p, q),$$

and the functions P, Q determine a canonical mapping **A**.

In fact, $p\,dq + Q\,dP$ is a closed form on R^{2n}; thus

$$dp \wedge dq = dP \wedge dQ .$$

(Q. E. D.)

Unfortunately, the generating function A is not a geometric object: A not only depends on the mapping A, but also on the coordinates p, q of R^{2n}.

According to (A 32.6), the generating function of the identity[1] is Pq. Thus, every canonical mapping, close enough to the identity, has a generating function close to Pq.

§B. Infinitesimal Canonical Mappings

Consider a family of canonical mappings S_ε, the generating functions $Pq + \varepsilon S(P, q; \varepsilon)$ of which depend smoothly on a parameter $\varepsilon \ll 1$. The mapping S_ε is close to the identity, if ε is small. According to (A 32.6), the Taylor expansions of $P(p, q)$ and $Q(p, q)$ with respect to ε are:

$$\text{(A 32.8)} \qquad P = p - \varepsilon \frac{\partial S}{\partial q} + O(\varepsilon^2), \qquad Q = q + \varepsilon \frac{\partial S}{\partial p} + O(\varepsilon^2) ,$$

where

$$S = S(p, q; \varepsilon) .$$

By definition, the infinitesimal canonical mapping S_ε is a class of equivalent families S_ε : two families S_ε and S_ε' are equivalent if $|S_\varepsilon - S_\varepsilon'| = O(\varepsilon^2)$.

DEFINITION A 32.9

The function $S(p, q)$ on the phase-space is called the generating function of the infinitesimal mapping S_ε (or Hamiltonian function).

Of course, S is defined up to a constant. Now we prove that the function S is a geometric object: S neither depends on the canonical coordinates p, q, nor on the choice of a representant S_ε in the class of equivalence: it is a mapping $S: R^{2n} \to R^1$. In fact, let γ be a curve of R^{2n}

---

[1] This is a way to memorize (A 32.6).

joining x and y: $\partial\gamma = y - x$. We set

$$\gamma_\varepsilon = S_\varepsilon \gamma ,$$

and we denote by $\sigma(\varepsilon)$ the strip formed by the curves $\gamma_{\varepsilon'}$, $0 < \varepsilon' < \varepsilon$, and oriented in such a way that $\partial\sigma_\varepsilon = \gamma - \gamma_\varepsilon + \cdots$ (see Figure A 32.10).

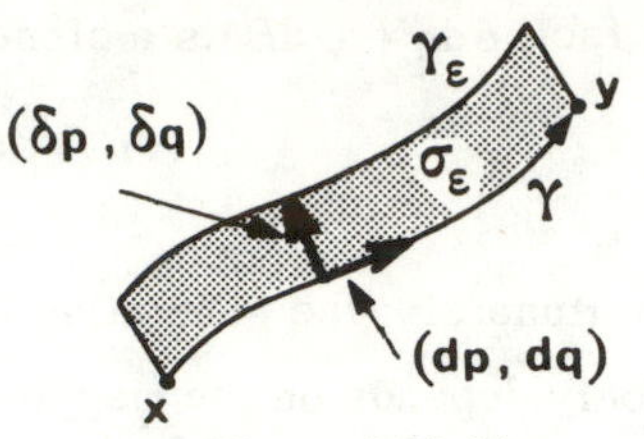

Figure A 32.10

Let us set:

$$I[\sigma(\varepsilon)] = \iint_{\sigma(\varepsilon)} dp \wedge dq . \tag{A 32.11}$$

According to (A 32.2), this integral does not depend on the canonical coordinates and, according to (A 32.1), does not depend on the curve γ, but only on x and y.

LEMMA A 32.12

The generating function S of the infinitesimal canonical mapping S_ε is given by:

$$S(y) - S(x) = \frac{d}{d\varepsilon} I[\sigma(\varepsilon)]\Big|_{\varepsilon=0} , \tag{A 32.13}$$

and does not depend on the choice of the canonical coordinates p, q.

Proof:

Let us set $S_\varepsilon x - x = \delta x = (\delta p, \delta q)$. According to (A 32.8), we have:

$$\varepsilon(S(y) - S(x)) = \varepsilon \int_\gamma \frac{\partial S}{\partial p} dp + \frac{\partial S}{\partial q} dq \tag{A 32.14}$$

$$= \int_\gamma (\delta q \, dp - \delta p \, dq) + O(\varepsilon^2) .$$

On the other hand, according to (A 32.11), the integral of $dp \wedge dq$ along $\sigma(\varepsilon)$ is:

$$I[\sigma(\varepsilon)] = \iint_{\sigma(\varepsilon)} dp \wedge dq = \int_\gamma \begin{vmatrix} dp & dq \\ \delta p & \delta q \end{vmatrix} + O(\varepsilon^2) . \tag{A 32.15}$$

Formulas (A 32.14) and (A 32.15) imply (A 32.13). (Q. E. D.)

One can express the invariance of the generating function S in another form. Let A be a finite canonical mapping and S_ε an infinitesimal canonical mapping. The canonical mapping $T_\varepsilon = AS_\varepsilon A^{-1}$ is clearly infinitesimal.

LEMMA A 32.16

The generating functions S and T of the infinitesimal mappings S_ε and T_ε are related by:

(A 32.17) $$T(Ax) = S(x) + \text{constant}.$$

Proof:

Let γ_ε and $\sigma(\varepsilon)$ be the curve and the surface of Lemma (A 32.12). The curve $\gamma' = A\gamma$ joins the points Ax and Ay. Besides, the curves $T_{\varepsilon'}\gamma'$, $0 \leq \varepsilon' \leq \varepsilon$, form a strip $\tau(\varepsilon)$, which is nothing but:

(A 32.18) $$\tau(\varepsilon) = A\sigma(\varepsilon).$$

From (A 32.13) follows:

(A 32.19) $$S(y) - S(x) = \frac{d}{d\varepsilon} I[\sigma(\varepsilon)], \quad T(Ay) - T(Ax) = \frac{d}{d\varepsilon} I[\tau(\varepsilon)].$$

But the mapping A is canonical. Thus, according to (A 32.2) and (A 32.18), we have:

$$I[\sigma(\varepsilon)] = I[\tau(\varepsilon)].$$

Comparison with (A 32.19) yields (A 32.17). (Q. E. D.)

COROLLARY A 32.20

Let B_ε and C_ε be infinitesimal canonical mappings with corresponding generating functions B and C, and let A be a finite canonical mapping. Then, the infinitesimal canonical mapping:

(A 32.21) $$B'_\varepsilon = C_\varepsilon B_\varepsilon A C_\varepsilon^{-1} A^{-1}$$

has the following generating function:

(A 32.22) $\quad B'(x) = C(x) + B(x) - C(A^{-1}x) + \text{constant.}$

In fact, (A 32.8) implies that the generating function of the product of two infinitesimal mappings is the sum of their generating functions, and also that the generating function of the inverse mapping C_ε^{-1} is $-C$. Relation (A 32.22) is easily derived from these remarks and from Lemma (A 32.16).

§C. Lie Commutators and Poisson Brackets

Given two infinitesimal canonical mappings A_ε and B_ε, there exists one and only one infinitesimal canonical mapping C_ε such that:

(A 32.23) $\quad A_a B_b A_{-a} B_{-b} = C_{ab} + O(a^2) + O(b^2)\ ; \quad a, b \to 0\ .$

The mapping C_ε is called the Lie commutator of A_ε and B_ε.

LEMMA A 32.24.

The generating function C of C_ε is equal, up to a sign, to the Poisson bracket of the generating functions A and B of A_ε and B_ε:

(A 32.25) $\quad \nabla C = -[\nabla A, \nabla B], \quad \nabla = \text{gradient.}$

We use the notation $[x, y] = (Ix, y)$ as in Appendixes 26 and 27.

Proof:

Again let γ be a curve joining x and y: $\partial\gamma = y - x$. We consider the five-sided prism (see Figure A 32.26) formed by four strips:

$$\sigma_1 = B_\varepsilon\gamma\ ,\ -b < \varepsilon < 0, \quad \partial\sigma_1 = \gamma - \gamma_1 + \cdots,$$

$$\sigma_2 = A_\varepsilon\gamma_1,\ -a < \varepsilon < 0, \quad \partial\sigma_2 = \gamma_1 - \gamma_2 + \cdots,$$

$$\sigma_3 = B_\varepsilon\gamma_2, \quad 0 < \varepsilon < b, \quad \partial\sigma_3 = \gamma_2 - \gamma_3 + \cdots,$$

$$\sigma_4 = A_\varepsilon\gamma_3, \quad 0 < \varepsilon < a, \quad \partial\sigma_4 = \gamma_3 - \gamma_4 + \cdots,$$

and closed by a fifth strip σ_5, formed by the segments joining the corresponding points of γ and γ_4, $\partial\sigma_5 = \gamma_4 - \gamma \cdots$. Finally, we denote by τ_x and τ_y the bases of our prism, such that the two-chain

$$\sigma_1 + \sigma_2 + \sigma_3 + \sigma_4 + \sigma_5 + \tau_y + \tau_x = \Sigma$$

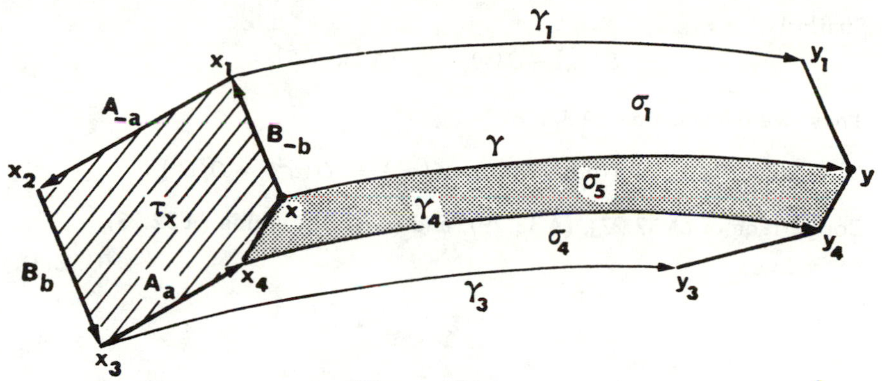

Figure A 32.26

forms a two-cycle homologous to zero. Since $dp \wedge dq$ is a closed form we have, with notations of (A 32.2):

(A 32.27)
$$I(\sigma_1) + I(\sigma_2) + I(\sigma_3) + I(\sigma_4) + I(\sigma_5) + I(\tau_y) + I(\tau_x) = I(\Sigma) = 0 .$$

But Lemma (A 32.12) implies:

(A 32.28)
$$\begin{cases} I(\sigma_1) = -b[B(y) - B(x)] + O(b^2), \\ I(\sigma_2) = -a[A(y_1) - A(x_1)] + O(a^2), \\ I(\sigma_3) = b[B(y_2) - B(x_2)] + O(b^2), \\ I(\sigma_4) = a[A(y_3) - A(x_3)] + O(a^2), \\ I(\sigma_5) = -ab[C(y) - C(x)] + O(a^2) + O(b^2). \end{cases}$$

On the other hand, $|y - y_4| = O(ab)$; therefore, the integrals of $\Sigma\, dp \wedge dq$ along the surfaces τ are given by:

(A 32.29)
$$\begin{cases} I(\tau_y) = -ab[\nabla B(y), \nabla A(y)] + O(a^2) + O(b^2), \\ I(\tau_x) = +ab[\nabla B(x), \nabla A(x)] + O(a^2) + O(b^2). \end{cases}$$

Finally, from (A 32.8) it follows that the vector fields corresponding to A_ε and B_ε are $I\nabla A$ and $I\nabla B$. Consequently, up to terms of order $O(a^2 + b^2)$, we have:

$$\begin{aligned} A(y_3) - A(y_1) &= (\nabla A, y_3 - y_1) = (\nabla A, y_3 - y_2) + (\nabla A, y_2 - y_1) \\ &= (\nabla A, I\nabla B)b - (\nabla A, I\nabla A)a = \\ &= [\nabla B, \nabla A]b - [\nabla A, \nabla A]a = [\nabla B, \nabla A]b . \end{aligned}$$

Similarly:

$$B(y_2) - B(y) = -[\nabla A, \nabla B]a .$$

Thus, we deduce from (A 32.28):

(A 32.30) $$I(\sigma_1) + I(\sigma_3) + I(\sigma_2) + I(\sigma_4) = O(a^2) + O(b^2) .$$

Comparison of (A 32.27), (A 32.29), and (A 32.30) yields (A 32.25).

(Q. E. D.)

APPENDIX 33

GLOBAL CANONICAL MAPPINGS

(See Section 21, Chapter 4)

This appendix gives the topological reasons for the existence of periodic orbits for Hamiltonian systems with n degrees of freedom.

§A. Generating Functions

Let $\Omega = B^n \times T^n$ be the canonical space, $p: \Omega \to B^n \subset \mathbf{R}^n = \{p_1, \ldots, p_n)\}$ and $q: \Omega \to T^n = \{(q_1, \ldots, q_n) \pmod{2\pi}\}$ the coordinates $p = p(x)$ and $q = q(x)$ of the point $x \in \Omega$. By definition, the mapping $\mathrm{A}: \Omega \to \Omega$ is *globally canonical* if it is homotopic to the identity and satisfies

(A 33.1)
$$\oint_{\gamma} p\mathrm{d}q = \oint_{\mathrm{A}\gamma} p\mathrm{d}q$$

for any one-cycle γ (even nonhomologous to zero). In conformity to Appendix 32, the mapping A is locally determined by a generating function $Pq + A(P, q)$, provided that

$$\mathrm{Det}\left(\frac{\partial P}{\partial p}\right) \neq 0,$$

(A 33.2)
$$p = P + \frac{\partial A}{\partial q}, \quad Q = q + \frac{\partial A}{\partial P}, \quad (\mathrm{A}x = (P(x), Q(x)),\ P \in B^n,\ Q \in T^n).$$

Thus, locally, the function $A(P, q)$ verifies:

(A 33.3)
$$A(P, q) = \int^{(P,q)} (Q - q)\mathrm{d}P + (p - P)\mathrm{d}q.$$

Let us set:

$$A(x) = A(P(x), q(x)) ,$$

where

$$x = (p(x), q(x)) \in \Omega .$$

LEMMA A 33.4

The mapping (A 33.2) *is globally canonical if and only if the function* $A(x)$, *defined by* (A 33.3), *is single-valued on* Ω .

Proof:

Let γ be a closed curve of Ω. Let us prove that:

(A 33.5) $$\oint_\gamma (Q-q)\mathrm{d}P + (p-P)\mathrm{d}q = 0 .$$

In fact, (A 33.1) is equivalent to:

(A 33.6) $$\oint_\gamma p\,\mathrm{d}q = \oint_\gamma P\,\mathrm{d}Q :$$

Thus, we obtain:

$$\oint_\gamma (Q-q)\mathrm{d}P + (p-P)\mathrm{d}q = \oint_\gamma Q\,\mathrm{d}P + P\,\mathrm{d}Q - (q\mathrm{d}P + P\,\mathrm{d}q) = \oint_\gamma \mathrm{d}[P(Q-q)].$$

But the increment of $P(Q-q)$ along γ is equal to zero, because $\mathbf{A}$ is homotopic to zero:

(A 33.7) $$\oint_\gamma \mathrm{d}[P\cdot(Q-q)] = 0 .$$

Conversely, (A 33.5) and (A 33.7) yield (A 33.6). (Q. E. D.)

§B. A Topological Lemma

Now, let $\mathbf{A}$ be a globally canonical diffeomorphism, T the torus, $p = 0$, and $\mathbf{A}T$ the image of T by $\mathbf{A}$.

LEMMA A 33.8

The tori T and AT have at least 2^n common points (counted with their order of multiplicity) provided that the equation of AT is:

$$p = p(q), \quad \left|\frac{dp}{dq}\right| < \infty . \tag{A 33.9}$$

Besides, the number of geometrically different points of intersection is, at least, $n + 1$.

Proof:

Consider the following function on AT:

$$f(x) = \int_{x_0}^{x} p(x)dq(x) , \tag{A 33.10}$$

where $x = (p(x), q(x)) \in AT$, and where the path of integration lies on AT. Function $f(x)$ is well-defined since the integral (A 33.10) does not depend on the path. In fact, let γ be a closed curve of AT, then:

$$\oint_{\gamma} pdq = \oint_{A^{-1}\gamma} pdq = 0,$$

because A^{-1} is globally canonical, $A^{-1}\gamma \subset T$, and $p = 0$ on T. The function $f(x)$ is a smooth function on the n-dimensional torus T^n. Thus, by the Morse[1] inequalities, the number of critical points is at least 2^n (and, following the Lusternic-Schnirelman theorem, the number of geometrically distinct points of intersection is at least the Lusternic-Schnirelman category of T^n, which is $n + 1$).

From (A 33.10) follows $df = p(x) \cdot dq(x)$ on AT. The function $p(x)$ vanishes at the intersection of AT and T. Thus, the points of intersection of T and AT are critical points of f on AT. Conversely, in view of condition (A 33.9), we have $pdq = 0$, for any dq, at every critical point x of f on AT; therefore $p(x) = 0$ and the critical point x belongs to the intersection of AT and T. (Q. E. D.)

[1] See Milnor [1]; $2^n = \sum_0^n b_i$, b_i = i-th Betti number of T^n.

COROLLARY A 33.11

The tori T and AT have at least 2^n common points, provided that their equations are:

$$p = p'(q), \quad p = p''(q) \left(\left|\frac{dp'}{dq}\right| < \infty, \quad \left|\frac{dp''}{dq}\right| < \infty \right) \tag{A 33.12}$$

and that $dp \wedge dq$ *vanishes on* T.

In fact, if $dp \wedge dq \equiv 0$ on T, then the mapping

$$p, q \to p - p'(q), q$$

is a canonical diffeomorphism. This diffeomorphism reduces (A 33.12) to (A 33.9) with

$$p(q) = p''(q) - p'(q) .$$

REMARK A 33.13

If $n = 1$ (mappings of annuli), Lemma (A 33.8) still holds without condition (A 33.9). The proof makes use of Jordan's theorem and does not extend for $n > 1$. Whether T and AT intersect, for $n > 1$, if condition (A 33.9) is not fulfilled, is an open question.

If condition (A 33.9) can be relaxed from Lemma (A 33.8), we obtain many "recurrence theorems" of the following type:

Assume that the initial values a_i, b_i of the axis of the Kepler ellipses, in the plane many-body problem, are such that the ellipses do not intersect. Then, whatever τ be, there exist initial phases[2] *l_i, g_i such that the axis of the ellipses return to their initial values after a time τ.*

REMARK A 33.14

If we drop condition (A 33.9), Lemma (A 33.8) cannot hold without assuming that A is a diffeomorphism, because (even for $n = 1$) regular and globally canonical mappings can be constructed such that T and AT do not intersect.

---

2 Phases l_i, g_i are angles (mod 2π); g_i determines the position of the ellipses and l_i determines the position of the planets on these ellipses.

§C. Fixed Points

Now, let A be a global canonical mapping of the following particular type:

(A 33.15) $\mathbf{A}: p, q \to p, q + \omega(p) \quad (\omega = (\omega_1, \ldots, \omega_n))$.

Assume that, on the torus $p = p_0$, all the frequencies are commensurate:

(A 33.16) $$\omega_i(p_0) = \frac{m_i}{N} 2\pi, \quad m_i \in \mathbf{Z}, \; N \in \mathbf{Z},$$

and that the twisting is nondegenerate:

(A 33.17) $$\mathrm{Det}\left(\frac{\partial \omega}{\partial p}\right)_{p_0} \neq 0 .$$

THEOREM A 33.18

Every globally canonical mapping B, C^1*-close enough to* A, *has at least* 2^n *points*[3] *with period* N *in the neighborhood of the torus* $p = p_0$:

$$\mathbf{B}^N x = x .$$

Proof:

In view of (A 33.15), (A 33.16), and (A 33.17) the mapping $\mathbf{A}^N$ can be written under the form:

(A 33.19) $\mathbf{A}^n$: $p, q \to p, q + \alpha(p)$, where $\alpha(p_0) = 0$, $\mathrm{Det}\left(\frac{\partial \alpha}{\partial p}\right)_{p_0} \neq 0$

in the neighborhood of the torus $p = p_0$. It is sufficient to put:

$$\alpha(p) = N[\omega(p) - \omega(p_0)] .$$

The neighboring mapping $\mathbf{B}^n$ can be written under the form:

(A 33.20) $\mathbf{B}^N$: $(p, q \to p + \beta_1(p, q), q + \alpha(p) + \beta_2(p, q)) = (P, Q)$.

---

[3] Counted with their multiplicity.

Let us consider the points which move along the radii: $Q = q$, that is:

(A 33.21) $$\alpha(p) + \beta_2(p, q) = 0 .$$

From the implicit function theorem we deduce:

(1) equation (A 33.21) determines a torus T, close to $p = p_0$ and which moves along the radii;

(2) the tori T and $\mathbf{B}^N T$ have equations of the form:

(A 33.22) $$p = p'(q), \quad p = p''(q), \text{ where } \left|\frac{dp'}{dq}\right| < \infty , \quad \left|\frac{dp''}{dq}\right| < \infty .$$

The mapping $\mathbf{B}^N$, being globally canonical, is given by a generating function of the form $Pq + B(P, q)$. In view of Lemma (A 33.4), the function $B(x) = B(P(x), q(x))$ is single valued in Ω. Now, consider the restriction of $B(x)$ to the torus T. This is a smooth function on an n-dimensional torus, thus B has at least 2^n critical points (compare to Lemma A 33.8).

Let us prove that these critical points belong to the intersection of T and $\mathbf{B}^N T$. Formula (A 33.3) yields:

$$dB = (Q-q)dP + (p-P)dq, \text{ where } \mathbf{B}: x = (p, q) \to (P(x), Q(x)) .$$

But, according to (A 33.20) and (A 33.21), we have $Q-q = 0$ on our torus T. Therefore $(p-P)dq = 0$ at the critical points of B on T. But, in view of (A 33.22), this implies $P = p$.

(Q. E. D.)

REMARK A 33.23

This theorem is not a corollary of Lemma (A 33.8): the manifold $Q-q = 0$ need not verify $dp \wedge dq = 0$, as the following example of canonical mapping proves:

$$P_1 = 3p_1 + 4p_2 + q_1 + q_2 , \qquad P_2 = p_1 + 3p_2 + q_2 ,$$

$$Q_1 = p_1 + p_2 + q_1 , \qquad Q_2 = p_2 - q_1 + q_2 .$$

APPENDIX 34

PROOF OF THE THEOREM ON THE CONSERVATION OF INVARIANT TORI UNDER SMALL PERTURBATIONS OF THE CANONICAL MAPPING

(See Theorem 21.11, Section 21, Chapter 4)

The construction of the invariant tori is performed in Sections E – H of this appendix. Lemmas of Sections B – D are used. Proof is based upon a method of successive approximations of Newtonian type suggested by A. N. Kolmogorov [6].

§A. Newton's Method

Newton's approximation for the construction of a zero x of a function f, with a prescribed approximation, consists in replacing the curve $y = f(x)$ by its tangent at $(x_0, f(x_0))$, where x_0 is an approximation of x. If $|x - x_0| < \varepsilon$, the substitution gives an error of order ε^2. Hence, the linearized equation

$$f(x_0) + f'(x_0)(x - x_0) = 0$$

admits a solution x_1 whose deviation from x will be of the order ε^2 (see Figure A 34.1). Iteration of this process leads to an accelerated convergent approximation:

(A 34.2) $$|x_{n+1} - x_n| < C|x_n - x_{n-1}|^2 \ .$$

Hence, the deviation of the nth approximation from the solution will be:

$$|x - x_n| \sim \varepsilon^{2^{n-1}} .$$

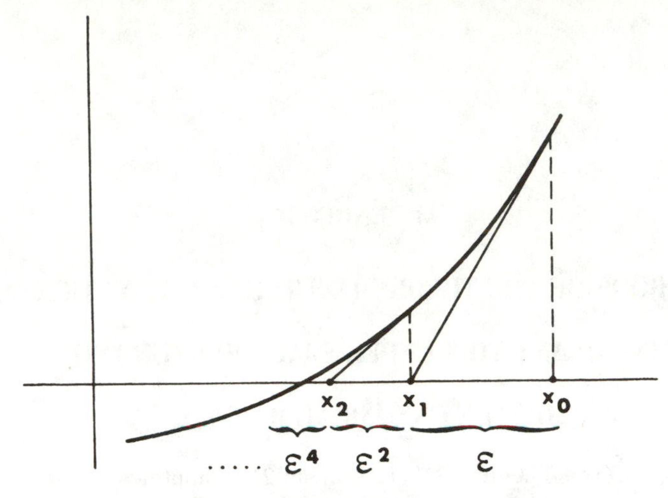

Figure A 34.1

This method extends easily to equations in Banach spaces[1]. But, in analysis, one is most often concerned with poly-Banachic spaces, and the operator $f'(x_0)^{-1}$ maps a space into another one. This situation is illustrated by the following lemma, which offers a characteristic example, and where (A 34.4) plays the role of (A 34.2).

LEMMA A 34.3[2]

Let L be an operator which maps the functions $f(z)$, which are analytic in each complex domain G, into the functions $Lf(z)$ which are analytic in the domain[3] *$G-\delta$ and such that, for every $0<\delta<\delta_0$:*

(A 34.4) $$|Lf|_{G-\delta} < |f|_G^2 \cdot \delta^{-\nu} ,$$

where $\nu > 0$, $\delta_0 > 0$ are absolute constants. Then, for every $\delta' > 0$, the series $\Sigma_s |L^s f|$ converges in $G-\delta'$, provided that $|f|_G < M = M(\delta')$ is small enough.

---

1 See Kantorovich [1].

2 This lemma was already known by H. Cartan [1], whose paper [1] is one of the first appearances of Newton's approximations in theoretical analysis.

3 Here and further, $G-\delta$ denotes the set of the points of G which are at a distance $>\delta$ from the complement of G. Example of an operator which verifies (A 34.4): $Lf = f \cdot \dfrac{df}{dz}$.

Proof:

Let

$$M_1 = \delta_1^{2\nu+1} \text{ and } \delta_2 = \delta_1^{3/2}, \ldots, \delta_{s+1} = \delta_s^{3/2}, \ldots,$$

$$M_2 = M_1^{3/2}, \ldots, M_{s+1} = M_s^{3/2},$$

hence

$$M_s = \delta_s^{2\nu+1} .$$

Then, if

$$\delta_1 < \frac{1}{8}, \frac{\delta'}{2},$$

we have:

$$\sum \delta_s < \delta', \qquad \sum M_s < 2M_1 .$$

Now, let

$$G_1 = G, \ G_2 = G_1 - \delta_1, \ldots, G_{s+1} = G_s - \delta_s .$$

Then, $|f|_G < M_1$ implies $|L^s f|_{G_s} < M_s$ because, in view of (A 34.4), $|f|_{G_s} < M_s$ implies:

$$|Lf|_{G_{s+1}} = |Lf|_{G_s - \delta_s} < M_s^2 \delta_s^{-\nu} < \delta_s^{3\nu+2} < \delta_{s+1}^{2\nu+1} = M_{s+1} .$$

But $G_s \supset G - \delta'$ for any s, because $\Sigma \delta_s < \delta'$. Therefore, we have:

$$|L^s f| < M_s, \ \Sigma |L^s f| < \Sigma M_s < 2M_1$$

in $G - \delta'$.

(Q. E. D.)

§B. Small Denominators

Let $f(q)$ be a function on a torus T^n, $q = (q_1, \ldots, q_n) \pmod{2\pi}$:

$$f(q) = \sum_{k \neq 0} f_k \cdot e^{i(k,q)} ,$$

where $(k, q) = k_1 q_1 + \cdots + k_n q_n$, and let $\omega = (\omega_1, \ldots, \omega_n)$ be a vector with irrational components, such that $(k, \omega) \neq k_0$ for nonvanishing integers

k, k_0. Consider the equation

(A 34.5) $$g(q+\omega) - g(q) = f(q),$$

where g is an unknown 2π periodic function. This equation admits a "formal" solution:

(A 34.6) $$g(q) = \sum_{k\neq 0} g_k \cdot e^{i(k,q)}, \qquad g_k = \frac{f_k}{e^{i(k,\omega)} - 1}.$$

The following lemma asserts the convergence of (A 34.6):

LEMMA A 34.7

Assume that the function $f(q)$ is analytic and $|f(q)| < M$ always holds for $|\mathrm{Im}\, q| < \rho$. Then, for almost every vector ω (except for a set of Lebesgue measure zero), the function $g(q)$, defined by (A 34.6), *is analytic and, for $|\mathrm{Im}\, q| < \rho - \delta$, we have: $|g(q)| < M\delta^{-\nu}$, $\nu = 2n+4$, if $0 < \delta < \delta_0$. Here $\delta_0 > 0$ is an absolute constant (independent of n).*

Proof of this lemma[4] is based on elementary results from the Theory of Diophantine approximations. In fact, Lemma (A 34.7) holds, with $\delta_0 = \delta_0(n, K)$ and for some $K > 0$, for any element ω of the set $\Omega_0(K)$ (defined below). Let us denote by $\Omega(K)$ the set of the ω_0 satisfying:

(A 34.8) $$|e^{i(k,\omega)} - 1| > KN^{-\nu}$$

for any ω, $|\omega - \omega_0| < KN^{-\nu}$, $\nu = n+2$, and for any k, $|k| \leq N$. Let us denote by $\Omega_0(K)$ the set of the ω_0 verifying:

(A 34.9) $$|e^{i(k,\omega_0)} - 1| > K|k|^{-\nu}, \qquad \nu = n+2.$$

Of course, $\Omega(K) \subset \Omega_0(K)$.

LEMMA A 34.10

Almost every (in the Lebesgue measure sense) point ω_0 belongs to $\Omega(K)$ for some $K > 0$ (hence, $\omega_0 \in \Omega_0(K)$).

---

[4] The technique of evaluating small denominators was extensively worked out by C. L. Siegel [2], [3], in connection with similar problems.

Proof:

Let Ω be a bounded domain of the space $\{\omega_0\}$. Let:

$$\Gamma_{k,d} = \{\omega_0 | \, |e^{i(k,\omega)} - 1| < d \text{ for some } \omega, \, |\omega - \omega_0| < d\} .$$

Then, clearly: $\text{meas}\,(\Gamma_{k,d} \cap \Omega) \leq C \cdot d$, where the constant C depends only on Ω. Relation (A 34.8) holds outside of $\bigcup_k \Gamma_{k,K|k|^{-\nu}}$. But we have:

$$\text{meas}\left(\bigcup_k \Gamma_{k,K|k|^{-\nu}} \right) \cap \Omega \leq \sum_k CK\,|k|^{-\nu} \leq C'K ,$$

since

$$\sum_k |k|^{-\nu} < \infty, \text{ for } \nu = n+2 .$$

Therefore:

$$\text{meas} \bigcap_{k \to 0} \overline{\Omega(K)} = 0 .$$

(Q. E. D.)

Now, let

$$f(q) = \sum_k f_k \cdot e^{i(k,q)}$$

be an analytic function.

LEMMA A 34.11

(A) If for $|\text{Im}\, q| < \rho$ we have $|f(q)| < M$, then $|f_k| < Me^{-\rho\,|k|}$.

(B) If $|f_k(q)| < Me^{-\rho\,|k|}$, then for $|\text{Im}\, q| < \rho - \delta$ (where $0 < \delta < \delta_0$):

$$|f(q)| < M\delta^{-\nu}, \qquad \nu = n+1 .$$

(C) If for $|\text{Im}\, q| < \rho$ we have $|f(q)| < M$, then for $|\text{Im}\, q| < \rho - \delta$, $0 < \delta < \delta_0$:

$$|R_N f| < Me^{-\delta N} \cdot \delta^{-\nu} .$$

Here, δ_0 and ν are absolute constants, which depend only on n, and R_N is:

$$R_N f = \sum_{|k| > N} f_n \cdot e^{i(k,q)} .$$

To prove (A) we have to shift the contour of integration in the formula:

$$f_k = \int f \cdot e^{-i(k,q)}\,dq$$

to $\pm\, i\rho$.

Proofs of (B) and (C) consist in mere summations of geometrical series.

Lemma (A 34.7) follows at once from Lemmas (A 34.10) and (A 34.11): take $\omega_0 \in \Omega_0(K)$ and take into account (A 34.6), (A 34.9), (A) and (B) from Lemma (A 34.11), and the elementary inequality:

$$e^{-|k|\delta} \cdot |k|^{\nu} < C(\nu)\delta^{-\nu} \quad .$$

Then, to obtain Lemma (A 34.7) it is sufficient to take $\delta_0 < K/C(\nu)$. We refer to Arnold [11] for further details.

REMARK A 34.12

Suppose that $\omega_0 \in \Omega(K)$ and $f_k = 0$ for $|k| > N$. Then, (A 34.6) reduces to a finite sum, which depends continuously on ω. Besides, Lemma (A 34.7) still holds, with the same $\delta_0 = \delta_0(K, n)$, for any ω' such that $|\omega - \omega'| < KN^{-\nu}$. Because if $\omega \in \Omega(K)$, then, in view of Definition (A 34.8), every ω', $|\omega - \omega'| < KN^{-\nu}$ verifies (A 34.9) for $|k| \leq N$. But the proof of Lemma (A 34.7) makes use of (A 34.9) only for $|k| \leq N$ if $f_k = 0$ for $|k| > N$.

§C. Sketch of the Proof

Now, let us recall notations of Theorem (21.11) (see Chapter 4). The set $\Omega = B^n \times T^n$ is a domain of the canonical space, a point x of Ω is denoted by $x = (p, q)$, where $p = (p_1, \ldots, p_n)$ is a point of the Euclidean ball B^n and $q = (q_1, \ldots, q_n) \pmod{2\pi}$ is a point of the torus T^n. The mapping $\mathbf{A}: p, q \to p, q + \omega(p)$ is the "unperturbed" mapping. The mapping $\mathbf{B}$, the generating function (see Appendix 32) of which is $Pq + B(P, q)$, is a small analytic globally canonical perturbation. We look for the invariant tori of the mapping $\mathbf{BA}$.

Following the idea of the Perturbation Theory (Appendix 30), we try to "kill" the perturbation B by an appropriate canonical coordinate transformation C, with generating function $Pq + C(P, q)$. In coordinate $\mathrm{C}x$, the mapping B A can be written:

$$\mathrm{C}(\mathrm{B\,A})\mathrm{C}^{-1} = \mathrm{B}'\mathrm{A}\ ,$$

where $\mathrm{B}' = \mathrm{C\,B\,A\,C}^{-1}\mathrm{A}^{-1}$. Therefore, taking into account Corollary (A 32.20), the generating function of B′ is:

$$Pq + B'(P, q), \text{ where } B'(x) = C(x) + B(x) - C(\mathrm{A}^{-1}x) + O(B^2 + C^2)\,.$$

Hence, killing B reduces to solving, with respect to C,

$$C(x) + B(x) - C(\mathrm{A}^{-1}x) = 0\,.$$

Now, observe that this equation, for each fixed p, has precisely the form of Equation (A 34.5) (with $\omega = \omega(p)$, $f = -B$, $g = C$). Now, constructing successive approximations to the invariant tori reduces to obtaining the inequalities:

(A 34.13) $$|B'|_{\Omega'} < |B|^2_{\Omega}\cdot\delta^{-\nu}\,, \quad |C|_{\Omega'} < |B|_{\Omega}\cdot\delta^{-\nu}$$

in a domain $\Omega' \subset \Omega$ (but not "too small"). Then, the convergence is carried out as in Lemma (A 34.3): Inequalities (A 34.13) are proved with the help of Lemma (A 34.7): they hold "far from resonances," that is for $\omega(p) \in \Omega(K)$.

To perform, along Sections E – H, the above program, we shall make use of several devices. First of all, instead of "killing" B, we only kill a truncation of its Fourier series: the remainder term $R_N B$ can be regarded as part of the "terms of higher order" $O(B^2)$ and $O(C^2)$. Hence, every approximation deals with only a finite number of resonances and small denominators in (A 34.6)[5]. On the other hand, to use Lemma (A 34.7), we

---

[5] This process was already used by Bogolubov, Mitropolski [1]. One could do without it by replacing the frequencies $\omega(p)$ with constants $\omega^* \in \Omega_K$ in the small denominators.

need first to eliminate the constant term B_0 (averaged value of B with respect to q): we add B_0 to the "unperturbed" mapping A (variation of the frequencies, Section E).

Finally, observe that, as pointed out by J. Moser [4], the Kolmogorov method still works in the case of differentiable mappings: in this case, one can devise a smoothing process like a suitable truncation of the Fourier series at the Nth frequency. Indeed, Moser [1] improved Theorem (21.11) in the case $n = 1$ (mappings of the plane) by abandoning the requirement of analyticity and substituting instead the requirement that 333 derivatives exist! Recently Moser [6] gave a proof which requires only a temperate number of derivatives.

§D. Canonical Mappings Close to the Identity

In this section we use the following notation: Let $F(\delta, M)$ be an arbitrary function and α a proposition involving δ, M, F. We shall say that *"α is true and $|F| \underset{\cap}{<} M$" if there exist absolute constants*[6] $\nu_1, \nu_2 > 0$ *and* $\delta_0 > 0$, *such that:*

(A 34.14) $$\alpha \text{ is true and } |F| \leq M\delta^{-\nu_2} ,$$

provided that $0 < \delta < \delta_0$, $M < \delta^{\nu_1}$.

LEMMA A 34.15

Let G be a complex domain and $f(z)$ an analytic function in G, which satisfies $|f(z)| < M$. Then, for $z \in G - \delta$, the k-th derivative of f is analytic and

$$\left|\frac{d^k f}{dz^k}\right| \underset{\cap}{<} M .$$

[6] That is that these constants depend on the dimensions of the domains dealing with the proposition α, on the number of derivatives, etc., but neither on the functions nor on the domains.

Proof:

From the Cauchy formula

$$\frac{d^k f}{dz^k} = \frac{k!}{2\pi i} \oint_\gamma \frac{F(\zeta)d\zeta}{(\zeta - z)^{k+1}}, \qquad \gamma : |\zeta - z| = \delta ,$$

we deduce:

$$\left| \frac{d^k f}{dz^k} \right| \leq k!\, M\delta^{-k} .$$

Hence, we obtain (A 34.14) with:

$$\nu_2 = k+1, \quad \delta_0 = \frac{1}{k!}, \quad \nu_1 = 0 .$$

(Q. E. D.)

Here:

$$F(\delta, M) = \sup_{f, G, z \in G-\delta} \left| \frac{d^k f}{dz^k} \right| ,$$

where sup ranges over all the bounded domains G, all the analytic functions f, $|f| < M$ in G, and all the points z of $G-\delta$.

Observe that: $CM\delta^{-\nu} \underset{\cap}{<} M$, and that the inequalities $F \underset{\cap}{<} M$ and $G \underset{\cap}{<} M$ imply $F + G \underset{\cap}{<} M$, $FG \underset{\cap}{<} M^2$. Then, if

$$F(\delta, M) \underset{\cap}{<} M ,$$

we have $F(C\delta, M) \underset{\cap}{<} M$ (here, C and ν are absolute positive constants).

Now, let us make clearer how the global canonical mappings S, close to the identity, are related to their generating functions $Pq + S(P, q)$. Let

$$\Omega = B^n \times T^n, \quad B^n = \{p \,\big|\, |p| < \gamma, \ p \in \mathrm{R}^n\} ,$$

$$T^n = \{q \ (\mathrm{mod}\ 2\pi), \ q = (q_1, \ldots, q_n)\} ,$$

and $[\Omega]$ be the complex domain of Ω, given by

$$|p| < \gamma, \quad |\mathrm{Im}\, q| < \rho, \text{ where } 0 < \gamma < 1, \ 0 < \rho < 1 .$$

LEMMA A 34.16

Let $S(p, q)$ be an analytic function in $[\Omega]$, satisfying:

$$|S(p, q)| < M .$$

Then, the formulas:

$$(A\,34.17) \qquad p = P + \frac{\partial S}{\partial q}, \quad Q = q + \frac{\partial S}{\partial P}, \quad S = S(P, q)$$

determine a global canonical diffeomorphism $\mathbf{S}$:

$$P = P(p, q), \quad Q = Q(p, q); \quad \mathbf{S}: \; [\Omega] - 2\delta \to [\Omega] - \delta,$$

and the following inequalities hold in $[\Omega] - 2\delta$:

$$\left|(P-p) - \frac{\partial S(p, q)}{\partial p}\right| \overset{<}{\cap} M^2, \quad \left|(Q-q) + \frac{\partial S(p, q)}{\partial q}\right| \overset{<}{\cap} M^2 .$$

LEMMA A 34.18

If $P = P(p, q)$, $Q = Q(p, q)$ *is a global analytic canonical mapping, which satisfies* $|P-p| < M$, $|Q-q| < M$ *in* $[\Omega]$, *then this mapping is defined in* $[\Omega]-\delta$ *by formulas* (A 34.17), *where* S *is analytic and verifies:*

$$|S| \overset{<}{\cap} M, \quad \left|S(p, q) - \int^{(p,q)} (Q-q)dp - (P-p)dq\right| \overset{<}{\cap} M^2 .$$

LEMMA A 34.19

Let $S(p, q)$ *and* $T(p, q)$ *be two analytic functions, which satisfy* $|S| < M$, $|T| < M$ *in* $[\Omega]$. *Then the product* $\mathbf{R} = \mathbf{ST}$ *of the corresponding canonical mappings is a global canonical diffeomorphism, which maps* $[\Omega]-3\delta$ *into* $[\Omega]-2\delta$, *and which is defined in* $[\Omega]-\delta$ *by an analytic generating function* R *satisfying:*

$$|R - (S+T)| \overset{<}{\cap} M^2 .$$

Now let $\mathbf{A}$: $[\Omega] \to [\Omega']$ be an analytic global canonical diffeomorphism ($[\Omega']$ is given by $|p| < \gamma'$, $|\mathrm{Im}\, q| < \rho'$, $0 < \gamma'$, $\rho' < 1$). Let $a^{-1}|y-x| < |\mathbf{A}(y) - \mathbf{A}(x)| < a|y-x|$, and S be an analytic function in $[\Omega]$, $\mathbf{S}$ the canonical diffeomorphism corresponding to S by virtue of Lemma (A 34.16).

LEMMA A 34.20

The formula $\mathrm{T} = \mathrm{ASA}^{-1}$ *defines a global canonical diffeomorphism of* $[\Omega'] - 3\delta$ *into* $[\Omega'] - 2\delta$, *which is given in* $[\Omega'] - \delta$ *by an analytic generating function* T *satisfying*[7]*:*

$$|T(\mathrm{A}x) - S(x)| \ \underset{\cap}{<}\ M^2 \ .$$

Proofs of the preceding lemmas reproduce those of Appendix 32, making use of Lemma (A 34.15) to compute the terms of order $O(\varepsilon^2)$. For further details we refer to Arnold [4], [5].

§E. Variation of the Frequencies

Let us now begin the construction of the invariant tori of the mapping BA (see Theorem 21.11, Chapter 4).

To understand the estimates performed in Sections E – F, it is useful to keep in mind that the positive numbers β, γ, δ, M and ρ, K, θ satisfy:

$$0 < M << \delta << \gamma << \beta << \rho,\ K,\ \theta^{-1} < 1 \ ,$$

and that ν_i, c_i are absolute constants, $\nu > 1 > c$.

CONSTRUCTION OF THE VARIATION OF THE FREQUENCIES. A 34.21

Let A and B be two global canonical mappings:

$$\mathrm{A}:\ p, q \to p, q + \omega(q) \ ,$$

and B have generating function $Pq + B(P, q)$. We put:

$$\bar{B}(p) = (2\pi)^{-n} \int \cdots \int B(p, q) dq_1 \cdots dq_n \ ,$$

$$\omega_1(p) = \omega(p) + \frac{\partial \bar{B}}{\partial p}$$

and we consider the following canonical mappings (see Figure A 34.22):

$$\mathrm{A}_1:\ p, q \to p, q + \omega_1(p) \ , \qquad \mathrm{B}' = \mathrm{B}\mathrm{A}\mathrm{A}_1^{-1} \ ,$$

---

[7] In this lemma, the constant δ_0, which enters into the definition of $\underset{\cap}{<}$, depends also on a.

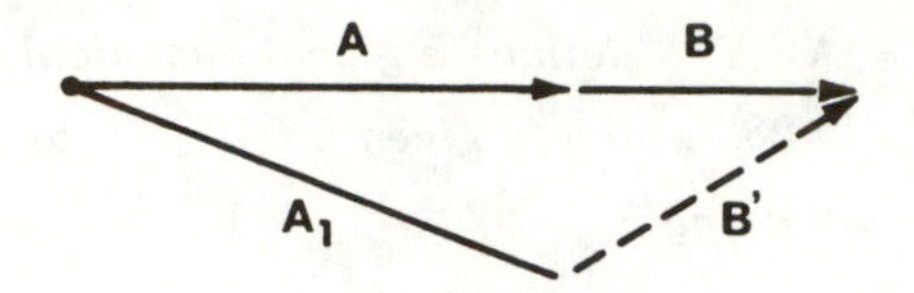

Figure A 34.22

Of course, $\mathbf{BA} = \mathbf{B}'\mathbf{A}_1$.

LEMMA CONCERNING THE VARIATION OF THE FREQUENCIES. A 34.23

Suppose that:

$$\theta^{-1}|dp| < |d\omega| < \theta|dp| \quad \text{and} \quad |B(p,q)| < M$$

hold in the domain $|p-p^*| < \gamma$, $|\text{Im}\, q| < \rho$. *Then the global canonical mapping* $\mathbf{B}'$ *possesses a generating function* $Pq + B'(P,q)$ *which satisfies*[8]:

$$|\omega_1(p) - \omega(p)| \underset{\cap}{<} M, \quad |B'| \underset{\cap}{<} M, \quad |\bar{B}'| \underset{\cap}{<} M^2,$$

where

$$\bar{B}' = (2\pi)^{-n} \int \cdots \int B'(p,q)dq_1 \cdots dq_n,$$

in the domain $|p-p^*| < \gamma-\delta$, $|\text{Im}\, q| < \rho-\delta$.

Proof:

Application of Lemma (A 34.19) to the mappings $\mathbf{B}$ and $\mathbf{A}\mathbf{A}_1^{-1}$: $p, q \to p, q - (\partial\bar{B}/\partial p)$ yields $|B' - (B-\bar{B})| \underset{\cap}{<} M^2$.

(Q. E. D.)

§F. Fundamental Lemma

We now make use of the estimates of the small denominators (Section B) to obtain inequalities of the form $|B_1| \underset{\cap}{<} M^2$, after a suitable change of coordinates $\mathbf{C}$.

---

[8] The constant δ_0, in the relation $\underset{\cap}{<}$ eventually depends on θ.

FUNDAMENTAL CONSTRUCTION. A 34.24

Let A and B be global canonical mappings:

$$\mathrm{A}:\ p, q \to p, q + \omega_1(p)\ ,$$

and B defined by a generating function $Pq + B(P, q)$,

$$B(P, q) = \bar{B}(P) + \sum_{k \neq 0} B_k(P) e^{i(k, q)}\ .$$

We put:

$$C(P, q) = \sum_{0 < |k| \leq N} C_k(P) e^{i(k, q)}\ ,$$

(A 34.25)

$$C_k(P) = \frac{B_k(P)}{e^{-i(k, \omega_1(P))} - 1}\ ,$$

where N is a positive integer. We denote by C the global canonical mapping, whose generating function is $Pq + C(P, q)$, and we consider the global canonical mapping $\mathrm{B}_1 = \mathrm{C B A C^{-1} A^{-1}}$. Of course:

$$\mathrm{B}_1\mathrm{A} = \mathrm{C(BA)C^{-1}}$$

(see Figure A 34.26). Observe also that:

$$\mathrm{A} \gg \mathrm{B} \sim \mathrm{C} \gg \mathrm{B}_1\ .$$

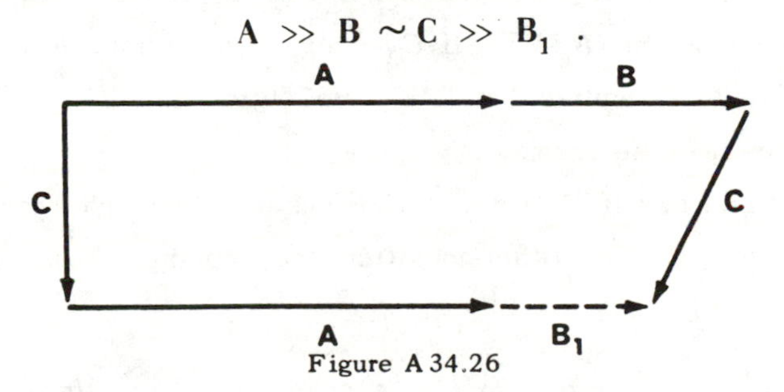

Figure A 34.26

FUNDAMENTAL LEMMA A 34.27

We suppose that in the domain $|p - p^*| < \gamma$, $|\mathrm{Im}\, q| < \rho$, *the functions* $\omega_1(p)$, $B(p, q)$ *are analytic and satisfy:*

$$\theta^{-1}|\mathrm{d}p| < |\mathrm{d}\omega_1(p)| < \theta|\mathrm{d}p|, \quad |B(p, q)| < M, \quad |\bar{B}(p)| < \bar{M}\ .$$

We suppose also that $\omega^* = \omega_1(p^*)$ *belongs to the set* $\Omega(K)$ *of Lemma* (A 34.10), *then:*

(1) *in the domain* $|P-p^*| < \gamma$, $|\mathrm{Im}\, q| < \rho-\delta$ *the function* $C(P, q)$ *is analytic and verifies:*

$$|C(P, q)| < M\delta^{-\nu_1} \ ;$$

(2) *in the domain* $|P-p^*| < \gamma-\delta$, $|\mathrm{Im}\, q| < \rho-\beta$, *the generating function* $Pq + B_1(P, q)$ *of* $\mathbf{B}_1$ *satisfies:*

$$(A\,34.28) \qquad |B_1(P, q)| < M^2\delta^{-\nu_1} + \bar{M} + M\cdot e^{-\beta N}\cdot\beta^{-\nu_1} \ ,$$

provided that [*for* (1) *and* (2)]:

$$(A\,34.29) \qquad \bar{M} < M < \delta^{\nu_2} \ ; \ \delta < C_1\gamma, \ \gamma < C_2\beta, \ \beta < C_3, \ \gamma < C_4 N^{-(n+2)} \ ,$$

where $\nu_1, \nu_2 > 1 > C_1, C_2, C_3, C_4 > 0$ *are absolute constants*[9].

Proof:

From $|d\omega| < \theta|dp|$ and $\gamma < C_4N^{-(n+2)}$, $C_4 < K/\theta$, it follows that all the points $\omega_1(p)$ of the domain $|p-p^*| < \gamma$ verify $|\omega-\omega^*| < KN^{-(n+2)}$. Consequently, according to the last remark of Section B, the assertion of Lemma (A 34.7) holds for the trigonometric sum C. Let $\delta_0 = \delta_0(n, K)$ be the constant of Lemma (A 34.7). If $C_3 < \delta_0$, then we have $\delta < \delta_0$; therefore Lemma (A 34.7) implies that $|C| < M\delta^{-\nu_1}$ for $|\mathrm{Im}\, q| < \rho-\delta$. This is precisely the (1) of our lemma.

Observe next that if C_2 (in $\gamma < C_2\beta$) is small enough, then we have $|\mathrm{Im}\, \omega(p)| < \theta\gamma < \beta$. Under this condition, the mappings $\mathbf{A}$ and $\mathbf{A}^{-1}$ are diffeomorphisms:

$$\{p, q \,|\, |p-p^*| < \gamma', \ |\mathrm{Im}\, q| < \rho'\} \to \{p, q \,|\, |p-p^*| < \gamma'', \ |\mathrm{Im}\, q| < \rho''\} \ ,$$

where $\rho'' < \rho' + \theta\gamma < \rho$. This allows one to apply Lemma (A 34.20) to

---

[9] That is, they depend only on the dimension n and the constants K, θ. The fundamental lemma states that there exist constants ν_1, ν_2 (large enough) and C_1, C_2, C_3, C_4 (small enough) such that (1) and (2) hold.

$AC^{-1}A^{-1}$. Besides, if ν_1 and ν_2 (in $M < \delta^{\nu_2}$) are large enough and C_3 (in $\delta < \beta < C_3$) small enough, the relations $\underset{\cap}{<}$ of Lemmas (A 34.16), (A 34.18), (A 34.19), and (A 34.20) are of the form: $< M\delta^{-\nu_1}$.

Hence, we deduce from these lemmas that, under the hypotheses of the fundamental lemma, with suitable constants C, ν, the mapping $B_1 = CBAC^{-1}A^{-1}$ has a generating function $Pq + B_1(P, q)$ such that in the domain $|P - p^*| < \gamma - \delta$, $|\mathrm{Im}\, q| < \rho - \beta$, the function $B_1(P, q)$ is analytic and:

(A 34.30) $$|B_1(x) - (B(x) + C(x) - C(A^{-1}x))| < M^2\delta^{-\nu_1} .$$

But from (A 34.25) follows:

(A 34.31) $$B(x) + C(x) - C(A^{-1}x) = \bar{B} + R_N B .$$

If $|\mathrm{Im}\, q| < \rho - \beta < \rho - \delta$, then assertion (C) of Lemma (A 34.11) implies:

(A 34.32) $$|R_N B| < Me^{-\beta N} \cdot \beta^{-\nu_1} .$$

Since $|\bar{B}| < \bar{M}$, formulas (A 34.30), (A 34.31), and (A 34.32) imply (A 34.28).

(Q. E. D.)

§G. The Inductive Lemma

The construction of the invariant tori makes use of an iterative process, each step being based on the following construction.

THE INDUCTIVE CONSTRUCTION. A 34.33

Let A and B be two canonical mappings:

$$A: \; p, q \to p, q + \omega(p) ,$$

B given by a generating function $Pq + B(P, q)$, and N a positive integer. Performing the variation of the frequencies (Section E) we obtain:

$$A_1: \; p, q \to p, q + \omega_1(p) ,$$

and B′, which is given by a generating function $Pq + B'(P, q)$ and such that $BA = B'A_1$. We now apply the fundamental construction of Section F to the mappings A_1, B′. We obtain a canonical mapping C, and $B_1 =$

$C B' A_1 C^{-1} A_1^{-1}$; hence (see Figure A 34.34) we have:

$$B_1 A_1 = C(BA)C^{-1} .$$

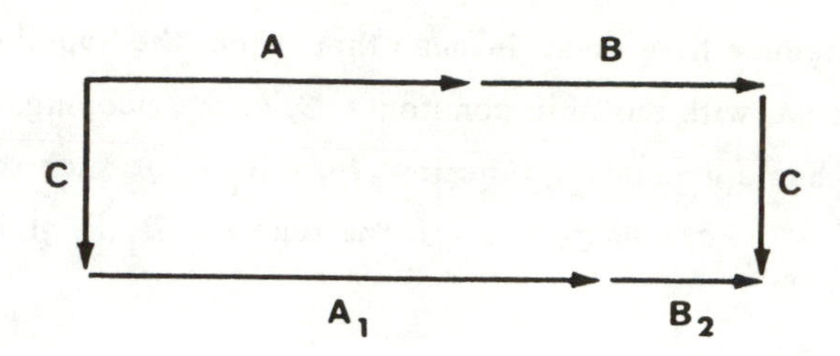

Figure A 34.34

THE INDUCTIVE LEMMA A 34.35

We suppose that in the domain $|p-p^*| < \gamma$, $|\mathrm{Im}\, q| < \rho$, *we have:*

$$\theta^{-1}|dp| < |d\omega| < \theta|dp|, \qquad \theta < \theta_0 ,$$

$$|B(p, q)| < M, \quad \omega(p^*) = \omega^* \in \Omega(K) .$$

Let us define p_1^* *by* $\omega_1(p_1^*) = \omega^*$, *and let* $Pq + B_1(P, q)$ *and* $Pq + C(P, q)$ *be the generating functions of* B_1 *and* C. *Then:*

(1) *in the domain* $|P-p^*| < \gamma$, $|\mathrm{Im}\, q| < \rho - \delta$, *the function* $C(P, q)$ *is analytic and* $|C| < M\delta^{-\nu_1}$;

(2) *the domain* $|P-p_1^*| < \gamma_1$. $|\mathrm{Im}\, q| < \rho_1 = \rho - \beta$ *belongs to the domain* $|P-p^*| < \gamma$, $|\mathrm{Im}\, q| < \rho$, *and in this domain the function* B_1 *is analytic and*

$$|B_1| < M^2\delta^{-\nu_4} + Me^{-\beta N}\beta^{-\nu_1} \;;$$

$$(3) \qquad \theta_1^{-1}|dp| < |d\omega_1| < \theta_1|dp|, \; |\theta_1 - \theta| < \delta, \; |\omega_1 - \omega| < \delta,$$

provided that:

$$M < \delta^{\nu_2};\; \delta < C_1\gamma, \gamma < C_2\beta,\; \beta < C_3;\; \gamma < C_4N^{-(n+2)},\; \gamma_1 < C_5\gamma .$$

where the constants $\nu_1, \nu_2, \nu_3, \nu_4 > 1 > C_1, C_2, C_3, C_4, C_5 > 0$ *are absolute constants, that is depend only on the dimension* n *and on the constants* θ_0, K, *but not on* B, ω, θ, M, *and so on.*

Proof:

Proof of the inductive lemma follows at once from the two preceding lemmas. The only two novelties are:

(1) *The existence of* p_1^* :

This existence and the inequality $|p_1^*| < C_5\gamma$ follow from the inequalities:

$$|d\omega_s| > \theta_0^{-1}|dp|, \quad |\omega_1 - \omega| < M\delta^{-\nu}, \qquad \left|\frac{d\omega_1}{dp} - \frac{d\omega}{dp}\right| < M\delta^{-\nu}$$

(see Section E) and from $M\delta^{-\nu} < C_6 \cdot \gamma$ (which follows from $M\delta^{-\nu} < \delta^{\nu_2 - \nu} < C_6 \cdot \gamma$ for ν_2 large enough). Taking into account $\gamma_1 < C_4\gamma$, the domain $|p - p_1^*| < \gamma_1$ belongs to $|p - p^*| < \gamma - \delta$ because $\delta < C_1\gamma$.

(2) *The estimate of* B_1 *:*

From Section E it follows that averaging B' gives:

$$|\overline{B}'| < M^2\delta^{-\nu} = \overline{M} .$$

Now replace $\overline{M}$ in relation (A 34.28) of the fundamental lemma, we obtain:

$$M^2\delta^{-\nu_1} + \overline{M} < M^2\delta^{-\nu_4} .$$

(Q. E. D.)

§H. Proof of Theorem (21.11) (see Chapter 4)

CONSTRUCTION A 34.36

The invariant torus $T(\omega^*)$ of the mapping BA, which corresponds to the frequencies ω^*, is constructed by making use of the process of the preceding section. This process depends on a sequence $0 < N_1 < N_2 < \cdots < N_s < \cdots, N_s \to +\infty$ which will be specified later.

After the sequence N_s is selected, the construction goes as follows. We put $\mathrm{A}_1 = \mathrm{A}$, $\mathrm{B}_1 = \mathrm{B}$, $N_1 = N$, and we make use of the inductive construction of Section G. This construction determines canonical mappings, namely C_1, A_2, B_2 ; we have:

$$\mathrm{B}_2\mathrm{A}_2 = \mathrm{C}_1(\mathrm{B}_1\mathrm{A}_1)\mathrm{C}_1^{-1} .$$

We again apply the same construction with $A_2 = A$, $B_2 = B$, $N_2 = N$; we obtain A_3, B_3, C_2, and so on. If A_s, B_s are constructed, the construction of Section G, with $A = A_s$, $B = B_s$, $N = N_s$ leads to C_s, A_{s+1}, B_{s+1} (see Figure A 34.37):

$$B_{s+1}A_{s+1} = C_s(B_sA_s)C_s^{-1} .$$

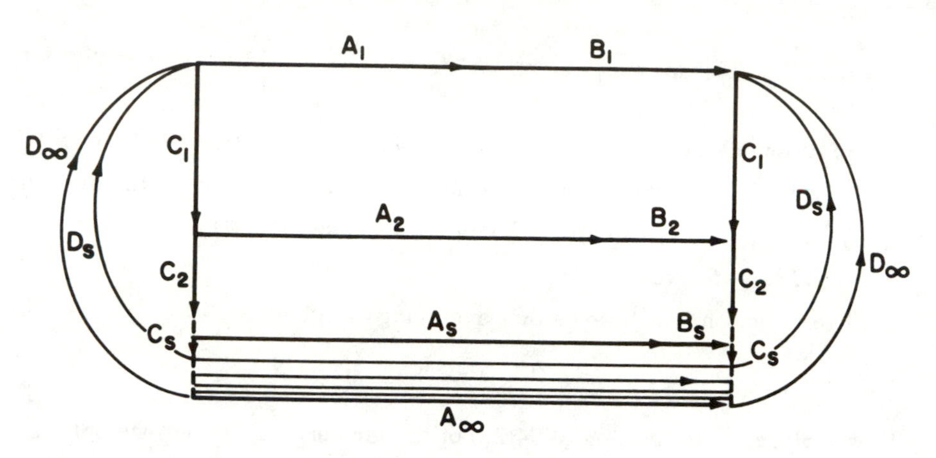

Figure A 34.37

This construction also determines the points p_s^*, $\omega_s(p_s^*) = \omega^*$. Now $T_s(\omega^*)$ denotes the torus $p = p_s^*$. The mapping A_s, restricted to this torus, is the translation defined by ω^*. We put $p_\infty^* = \lim_{s\to\infty} p_s$, and let T_∞^* be the torus $p = p_\infty^*$, A_∞ the mapping:

$$A_\infty: \ T_\infty^* \to T_\infty^*, \quad A_\infty q = q + \omega^* .$$

Finally, we put:

$$D_s = C_1^{-1} C_2^{-1} \cdots C_{s-1}^{-1}, \ D = \lim_{s\to\infty} D_s .$$

The invariant torus of BA is given by:

$$T(\omega^*) = DT_\infty^* .$$

We now prove that the above limits exist and that $DA_\infty = BAD$ on $T(\omega^*)$.

(Q. E. D.)

CONVERGENCE, A 34.38

In view of $d\omega/dp \not\equiv 0$, we can suppose that:

$$\theta^{-1}|dp| < |d\omega(p)| < \theta|dp|$$

holds in $[\Omega]$. Let us suppose that $\omega^* \in \Omega(K)$. According to Lemma (A 34.10), almost every ω^* belongs to $\Omega(K)$ for some $K > 0$. We put $\theta_0 = 2\theta$, and we define a sequence of constants by:

$$\delta_1 = \delta > 0, \ \delta_2 = \delta_1^{3/2}, \ \delta_3 = \delta_2^{3/2}, \ldots, \delta_{s+1} = \delta_s^{3/2}, \ldots .$$

Let us put:

$$\gamma_s = \delta_s^{1/2}, \ \beta_s = \gamma_s^{1/4(n+2)}, \ \ N_s = \beta_s^{-2} = \gamma_s^{-1/2(n+2)} \ ;$$

then we have:

$$\gamma_{s+1} = \gamma_s^{3/2}, \quad \beta_{s+1} = \beta_s^{3/2}, \quad N_{s+1} = N_s^{3/2} \ .$$

To define all these numbers, we just need to choose δ. Denote by α a positive constant, if α is taken large enough we have:

$$\alpha > \nu_2, \quad \alpha > 2\nu_4 + 1, \quad \alpha > \nu_1 + 2 \ .$$

Let C, $0 < C < 1/10$, be an absolute constant, that is, C depends only on n, K, θ_0, α and on the constants ν_k, C_k which enter into the inductive lemma. If $\delta < C$, then, obviously:

(1)

$$\Sigma \beta_s < \frac{1}{10} \rho \ ; \tag{A 34.39}$$

(2)

$$\Sigma \delta_s < \theta \ ; \tag{A 34.40}$$

(3) for any $s = 1, 2, \ldots$, we have:

$$\delta_s < C_1\gamma_s, \ \gamma_s < C_2\beta_s, \ \beta_s < C_3, \ \gamma_s < C_4 N_s^{-(n+2)}, \tag{A 34.41}$$

$$\gamma_{s+1} < C_5\gamma_1 .$$

where C_1, C_2, C_3, C_4, C_5 are the constants of the inductive lemma, which depend on the constants K and θ_0 we introduced above;

(4)

(A 34.42) $$e^{-\beta_s N_s} = e^{-1/\beta_s} < \delta_s^{(a/2+\nu_1)+1} .$$

Our number δ is now selected to satisfy: $0 < \delta < C$. Hence, the numbers N_s, on which our construction depends, are determined.

Suppose now that our mapping $\mathbf{B}_1 = \mathbf{B}$ has a generating function $Pq + B_1(P, q)$ which verifies $|B_1| < M_1 = \delta_1^{\,a}$ in $|p - p_1^*| < \gamma_1$, $|\mathrm{Im}\, q| < \rho_1 = \frac{1}{2}\rho$. According to Lemma (A 34.18) this inequality holds if ε, under the conditions of Theorem (21.11), is small enough. Taking into account inequalities (A 34.41), the hypotheses of the inductive lemma are satisfied by $\mathbf{A} = \mathbf{A}_1$, $\mathbf{B} = \mathbf{B}_1$ (because $a > \nu_2$). Therefore we obtain $\mathbf{A}_2$, $\mathbf{B}_2$, $\mathbf{C}_1$, p_2^*, and so on from the inductive lemma.

Let us prove that the mappings $\mathbf{B}_2$, $\mathbf{A}_2$ again satisfy all the conditions of the inductive lemma in the domain $|p - p_2^*| < \gamma_2$, $|\mathrm{Im}\, q| < \rho_2 = \rho_1 - \beta_1$. In fact, (A 34.39) implies $\rho_2 > 0$; (A 34.40) and the third part of the inductive lemma show that $\mathbf{A}_2$ satisfies the inequalities:

$$\theta_2^{-1}|dp| < |d\omega_2(p)| < \theta_2 \cdot |dp| \text{ with } \theta_2 < \theta_0 .$$

Finally, the second part of the inductive lemma, (A 34.42), and $a > 2\nu_4 + 1$, $a > \nu_1 + 2$, imply:

$$|B_2| < M_1^2 \delta_1^{-\gamma_4} + M_1 e^{-\beta_1 N_1} \cdot \delta^{-\nu_1} < \delta_1^{2a - \nu_4} + \delta_1^{3/2(a+1)} < \delta_1^{3a/2} .$$

In other words:

$$|B_2| < M_2 = \delta_2^{a} .$$

Hence, $\mathbf{B}_2$ and $\mathbf{A}_2$ fulfill all the conditions of the inductive lemma.

Iteration of this argument leads to:

$$|B_s| < M_s = \delta_s^{a} , \text{ for } |p - p_s^*| < \gamma_s, \ |\mathrm{Im}\, q| < \rho_s .$$

Let G_s be the domain $|p - p_s^*| < \gamma_s$, $|\mathrm{Im}\, q| < \rho_s$. Then, the diffeomorphisms $\mathbf{C}_s^{-1}$ map G_{s+1} into G_s and, in the C^1-norm, we have:

(A 34.43) $$\|\mathbf{C}_s^{-1} - E\|_{C_1} < \delta_s, \qquad E = \text{identity},$$

in G_{s+1}. The point p_∞^* is the intersection of the balls:

$$|p - p_s^*| < \gamma_s, \qquad s \to \infty .$$

From the estimate (A 34.43), follows at once the convergence of the $\mathbf{D}_s$ on the torus

$$T_\infty^* = \bigcap_{s \geq 1} G_s .$$

The inequalities $|B_s| < \delta_s^a$ imply:

$$|\mathbf{D}_s^{-1}\mathbf{B}\mathbf{A}\mathbf{D}_s - \mathbf{A}_s| = |\mathbf{B}_s\mathbf{A}_s - \mathbf{A}_s| \to 0 \text{ as } s \to \infty ,$$

on the torus T_∞^*.

Finally, from $|\omega_{s+1} - \omega_s| < \delta_s$ follows the convergence of the mappings $\mathbf{A}_s$:

$$\lim_{s \to \infty} \mathbf{A}_s = \mathbf{A}_\infty : \; q \to q + \omega^*, \text{ on } T_\infty^* .$$

Thus:

$$\mathbf{D}^{-1}(\mathbf{B}\mathbf{A})\mathbf{D} = \mathbf{A}_\infty, \text{ on } T_\infty^* .$$

(Q. E. D.)

BIBLIOGRAPHY

Abramov, L. M.

[1] Concerning a note of Genis, *R Z Math.* 8(1963) p. 439.

[2] Metric Automorphisms with Quasi-Discrete Spectrum, *Isvestia Math. Series* 26 (1962) pp. 513-531.

Adler, R. L., and Rivlin, T. V.

[1] Ergodic and Mixing Properties of Chebyschev Polynomials, *Proc. Amer. Math. Soc.* 15 No. 5 (1964) pp. 794-796.

Andronov, A. A., and Pontrjagin, L. S.

[1] Rough Systems, *Dokl. Akad. Nauk.* 14 (1937) p. 247-250.

Anosov, D. V.

[1] Roughness of Geodesic Flows on Compact Riemannian Manifolds of Negative Curvature, *Dokl. Akad. Nauk.* 145 (1962) p. 707-709. [*Sov. Math. Dokl.* 3 No. 4 (1962) pp. 1068-1069.]

[2] Ergodic Properties of Geodesic Flows on Closed Riemannian Manifolds of Negative Curvature, *Dokl. Akad. Nauk.* 151 (1963) pp. 1250-1253. [*Sov. Math. Dokl.* 4 No. 4 (1963) pp. 1153-1156.]

[3] Averaging in Systems of Ordinary Differential Equations with Rapidly Oscillating Solutions, *Isvestia, Math Series* 24 (1960) pp. 721-742.

Anzai, H.

[1] Ergodic Skew Product Transformations on the Torus, *Osaka Math. J.* (1951) pp. 83-99.

Arnold, V. I.

[1] Sur la géométrie des groupes de difféomorphismes et ses applications en hydrodynamique des fluides parfaits, *Ann. Inst. Fourier* (1966).

[2] Remarks on Rotation Numbers, *Sibirski Math. Zh.* 2 No. 6 (1961) pp. 807-813.

[3] Some Remarks on Flows of Line Elements and Frames, *Dokl. Akad. Nauk.* 138 No. 2 (1961) pp. 255-257. [*Sov. Math. Dokl.* 2 (1961) pp. 562-564.]

[4] Small Denominators and Problems of Stability of Motion in Classical and Celestial Mechanics, *Usp. Math. Nauk.* 18 No. 6 (1963) pp. 91-196. [*Russian Math. Surveys* 18 No. 6 (1963) pp. 85-193.]

[5] Proof of a Theorem of A. N. Kolmogorov on the Invariance of Quasi-Periodic Motions under Small Perturbations of the Hamiltonian, *Usp. Math. Nauk.* 18 No. 5 (1963) pp. 13-40. [*Russian Math. Surveys* 18 No. 5 (1963) pp. 9-36.]

Arnold, V. I., and Sinai, Y.

[6] Small Perturbations of the Automorphisms of a Torus, *Dokl. Akad. Nauk.* 144 No. 4 (1962) pp. 695-698. [*Sov. Math. Dokl.* 3 (1962) pp. 783-786.]

Arnold, V. I.

[7] On the Stability of Positions of Equilibrium of a Hamiltonian System of Ordinary Differential Equations in the General Elliptic Case, *Dokl. Akad. Nauk.* 137 No. 2 (1961) pp. 255-257. [*Sov. Math Dokl.* 2 pp. 247-279.]

[8] On the Generation of a Quasi-Periodic Motion from a Set of Periodic Motions, *Dokl. Akad. Nauk.* 138 No. 1 (1961) pp. 13-15. [*Sov. Math. Dokl.* No. 2 pp. 501-503.]

[9] On the Behavior of an Adiabatic Invariant under a Slow Periodic Change of the Hamiltonian, *Dokl. Akad. Nauk.* 142 No. 4 (1962) pp. 758-761. [*Sov. Math. Dokl.* No. 3, pp. 136-139.]

[10] On the Classical Theory of Perturbations and the Problem of Stability of Planetary Systems, *Dokl. Akad. Nauk.* 145 No. 3 (1962) pp. 487-490. [*Sov. Math. Dokl.* No. 3 (1962) pp. 1008-1011.]

[11] Small Denominators I, on the Mapping of a Circle into Itself, *Izvestia Akad. Nauk.* Math. Series 25, 1 (1961) pp. 21-86. [*Transl. Amer. Math. Soc.*, Series 2, 46 (1965) pp. 213-284.]

[12] Conditions for the Applicability, and Estimate of the Error, of an Averaging Method for Systems which Pass through States of Resonance in the Course of Their Evolution, *Dokl. Akad. Nauk.* 161 No. 1 (1965) pp. 9-12. [*Sov. Math. Dokl.* 6 No. 2 (1965) pp. 331-337.]

[13] Instability of Dynamical Systems with Many Degrees of Freedom, *Dokl. Akad. Nauk.* 156 No. 1 (1964) pp. 9-12. [*Sov. Math. Dokl.* 5 No. 3 (1964) pp. 581-585.]

[14] Stability and Instability in Classical Mechanics, *Second Summer Math. School,* 1964 2, Kiev (1965) p. 110.

Artin, M., and Mazur, B.

[1] On Periodic Points, *Ann. Math.* 81 No. 1 (1965) pp. 82-99.

Auslander, L., Green, L., and Hahn, F.
[1] Flows on Homogeneous Spaces, *Ann. Math. Studies* 53 (1963).

Avez, A.
[1] Quelques inégalités de géométrie différentielle globale déduites de la théorie ergodique, *C. R. Acad. Sci. Paris* t. 261 (1965) pp. 2274-2277.
[2] Spectre discret des systèmes ergodiques classiques, *C. R. Acad. Sci. Paris* t. 264 (1967) pp. 49-52.

Birkhoff, G. D.
[1] Dynamical Systems, New York (1927).

Blum, J. R. and Hanson D. L.
[1] On the Isomorphism Problem for Bernouilli Schemes, *Bull. Am. Math. Soc.* 69 No. 2 (1963) pp. 221-223.

Bogolubov, N. N. and Mitropolski, Y. A.
[1] Les méthodes asymptotiques en théorie des oscillations non linéaires, Gauthier-Villars, Paris (1962).

Bogolubov, N. N.
[2] Proceedings of a Summer Math. School at Kanev (1963) Editions "Naukova Dumka," Kiev (1964).

Bohl, P.
[1] Uber ein in der Theorie der Säkularen störungen vorkommendes Problem, *J. Reine u. Angew. Math.* 135 (1909) pp. 189-283.

Cairns, S. S.
[1] On the Triangulation of Regular Loci, *Ann. Math.* 35 No. 2 (1934) pp. 579-587.

Callahan, F. P.
[1] Density and Uniform Density, *Proc. Am. Math. Soc.* 15 No. 5 (1964) pp. 841-843.

Cartan, H.
[1] Sur les matrices holomorphes de *n* variables complexes, *J. Math. pures appli.* 19 (1940) pp. 1-26.

Chacon, R. V.
[1] Change of Velocity in Flows, *J. Math. Mech.* 16 No. 5 (1966).

Coddington, K. and Levinson, N.

[1] Theory of Ordinary Differential Equations, McGraw-Hill, New York (1955).

De Baggis, H. F.

[1] Dynamical Systems with Stable Structure. Contribution to the Theory of Nonlinear Oscillations, Vol. 2, *Ann. Math. Studies* 29 (1952) pp. 37-59.

Denjoy, A.

[1] Sur des courbes définies par des équations différentielles à la surface du tore, *J. de Math.* 9 (1932) pp. 333-375.

Euler, L.

[1] Theoria Motus Corporum Solidorum Seu Rigidorum (1765).

Gelfand, I. M. and Shapiro-Piatetzski, I. I.

[1] A Theorem of Poincaré, *Dokl. Akad. Nauk.* 127 No. 3 (1959) pp. 490-493. [*Math. Review* (1960) No. 6460.]

Gelfand, I. M. and Fomin, S. V.

[2] Geodesic Flow on Manifold of Constant Negative Curvature, *Usp. Math. Nauk.* 47 No. 1 (1952) pp. 118-137. [*Transl. Amer. Math. Soc.* 2 No. 1 (1955) pp. 49-67.]

Gelfand, I. M., Graev, M. I., Zueva, H. M., Michailova, M. S., and Morosov, A. I.

[3] An Example of a Toroidal Magnetic Field Not Having Magnetic Surfaces, *Dokl. Akad. Nauk.* 143 No. 1 (1962) pp. 81-83. [*Sov. Phys. Dokl.* 7 pp. 223-224.]

Gelfand, I. M. and Lidsky, V. B.

[4] On the Structure of Regions of Stability of Linear Canonical Systems of Differential Equations with Periodic Coefficients, *Usp. Math. Nauk.* 10 No. 1 (63) (1955) pp. 3-40. [*Transl. Amer. Math. Soc.* 2 (8) (1958) pp. 143-181.]

Genis, A. L.

[1] Metric Properties of the Endomorphisms of the n-Dimensional Torus, *Dokl. Akad. Nauk.* 138 (1961) pp. 991-993. [*Sov. Math. Dokl.* 2 (1961) pp. 750-752.]

Gourevitch, B. M.

[1] The Entropy of Horocycle Flows, *Dokl. Akad. Nauk.* 136 No. 4 (1961) pp. 768-770. [*Sov. Math. Dokl.* 2 (1961) pp. 124-130.

Grant, A.

[1] Surfaces of Negative Curvature and Permanent Regional Transitivity, *Duke Math. J.* No. 5 (1939) pp. 207-229.

Guirsanov, I. V.

[1] On the Spectra of Dynamical Systems Which Arise from Stationary Gaussian Process, *Dokl. Akad. Nauk.* 119 No. 5 (1958) pp. 851-853.

Gysin, W.

[1] Zur Homologietheorie der Abbildungen und Faserungen von Mannigfaltigkeiten, *Comment. Math. Helv.* 14 (1941) pp. 61-122.

Hadamard, J.

[1] Les surfaces a courbures opposées et leurs lignes geodésiques, *J. Math. pures appl.* (1898) pp. 27-73.

Hajian, A. B.

[1] On Ergodic Measure Preserving Transformations Defined on an Infinite Measure Space, *Proc. Am. Math. Soc.* 16 (1965) pp. 45-48.

Halmos, P. R.

[1] Lectures on Ergodic Theory, Chelsea, New York (1958).

[2] Measure Theory, New York (1951).

[3] Introduction to Hilbert Spaces, Chelsea, New York (1957).

Hedlund, G.

[1] The Dynamics of Geodesic Flows, *Bull. Am. Math. Soc.* 45 (1939) pp. 241-246.

Helgason, S.

[1] Differential Geometry and Symmetric Spaces, Academic Press, New York (1962).

Hénon, M. and Heiles, C.

[1] The Applicability of the Third Integral of Motion: Some Numerical Experiments, *Astron. J.* 69 No. 1 (1964) pp. 73-79.

Hopf, E.

[1] Ergodentheorie, Springer, Berlin (1937).

Iusvinskii, S. A.

[1] On Metrical Automorphisms with Simple Spectrum, *Dokl. Akad. Nauk.* 172 No. 5 (1967).

Jacoubovitch, V. I.

[1] Questions of the Stability of Solutions of a System of Linear Differential Equations of Canonical Form with Periodic Coefficients, *Mat. Sbornik.* 37, 79 (1955) pp. 21-68. [*Transl. Am. Math. Soc.* 2 10 (1958) pp. 125-175.]

Kagan, V. F.

[1] Foundations of the Theory of Surfaces, *Osnovy Teorii Poverchnosteï* 1, Moscow (1947).

Kantorovitch, L. B.

[1] Functional Analysis and Applied Mathematics, *Usp. Math. Nauk.* 3 No. 6 (1948) pp. 89-185.

Kasuga, T.

[1] On the Adiabatic Theorem for the Hamiltonian System of Differential Equations in the Classical Mechanics, I. II. III., *Proc. Japan. Acad.* 37 No. 7 (1961) pp. 366-382.

Katok, A. B.

[1] Entropy and Approximation of Dynamical Systems by Periodical Mappings, *Funkzionalnyi Analys i ego Prilojenija,* Moscow, 1, No. 1 (1967) pp. 75-85.

[2] On Dynamical Systems with Integral Invariants on the Torus, *Funkz. Anal. i ego Prilojenija,* Moscow 1 No. 3 (1967).

Katok, A. B. and Stepin, A. M.

[1] On the Approximations of Ergodic Dynamical Systems by Periodical Mappings, *Dokl. Akad. Nauk.* 171 No. 6 (1966) pp. 1268-1271.

Kolmogorov, A. N.

[1] Sur les systemes dynamiques avec un invariant intégral à la surface du tore, *Dokl. Akad. Nauk.* 93 No. 5 (1953) pp. 763-766.

[2] A New Metric Invariant of Transitive Systems and Automorphisms of Lebesgue Spaces, *Dokl. Akad. Nauk.* 119 (1958) pp. 861-864. [*Math. Review* 21 No. 2035a.]

[3] Foundations of Probability Theory, Chelsea, New York (1956).

[4] On the Entropy per Time Unit as a Metric Invariant of Automorphisms, *Dokl. Akad. Nauk.* 124 (1959) pp. 754-755.

[5] La théorie générale des systèmes dynamiques et la mécanique classique, Amsterdam Congress I (1954) pp. 315-333. [*Math. Review* 20 No. 4066.]

[6] On the Conservation of Quasi-Periodic Motions for a Small Change in the Hamiltonian Function, *Dokl. Akad. Nauk.* 98 No. 4 (1954)

pp. 527-530. [*Math. Review* 16 No. 924.]
[7] Lectures given in Paris (1956).

Koopman, B. O.
[1] Hamiltonian Systems and Transformations in Hilbert Spaces, *Proc. Natl. Acad. Sci. U.S.* 17 (1931) pp. 315-318.

Kouchnirenko, A. G.
[1] An Estimate from Above for the Entropy of a Classical System, *Dokl. Akad. Nauk.* 161 No. 1 (1965) pp. 37-38. [*Sov. Math. Dokl.* 6 No. 2 (1965) pp. 360-362.]
[2] Every Analytical Action of a Semi-Simple Lie Group in the Neighborhood of a Fixed Point is Equivalent to the Linear Action, *Funkz. Analys i ego Prilojenija*, Moscow, 1 No. 1 (1967) pp. 103-104.
[3] Sur les invariants métriques du type entropie (Russian) Int. Congress, Moscow VIII (1966).

Krasinskii, G. A.
[1] Normalization of a System of Canonical Differential Equations Near a Quasi-Periodic Motion, *Bull. Inst. Theor. Astr.*, Leningrad (1967).

Krein, M. G.
[1] A Generalization of Several Investigations of A. M. Lyapounov, *Dokl. Akad. Nauk.* 73 (1950) pp. 445-448. [*Math. Review* 12 No. 100.]
[2] The Basic Propositions of the Theory of λ-Zones of Stability of a Canonical System of Linear Differential Equations with Periodic Coefficients, Pamyati A. A. Andronova, *Izvestia Akad. Nauk.* (1955) pp. 413-498. [*Math Review* 17 No. 738.]

Kupka, I.
[1] Stabilité des variétés invariantes d'un champ de vecteurs pour les petites perturbations, *C. R. Acad. Sci. Paris* 258 (1964) pp. 4197-4200.

Lagrange, R.
[1] Oeuvres t. 5, pp. 123-344.

Leontovitch, A. M.
[1] On the Stability of the Lagrange Periodic Solutions for the Reduced Problem of the Three Bodies, *Dokl. Akad. Nauk.* 143 No. 3 (1962) pp. 525-528. [*Sov. Math. Dokl.* 3 No. 2 (1962) pp. 425-430.]

Levi-Civita, T.
[1] Sopra alcuni criteri di instabilita, *Ann. Mat. pura appl.* 3, 5 (1901) pp. 221-307.

Margoulis, A.

[1] Sur quelques questions Liées à la théorie des C-systèmes d'Anosov (Russian) Int. Congress, Moscow VIII (1966).

Melnikov, V. K.

[1] On the Stability of a Center for Time Periodic Perturbations, *Trudy Moskovskogo Math. Obschestva* 12 No. 3 (1963) pp. 3-53. [*Math. Review* No. 5981 (1964).]

[2] On Some Case of Conservation of Conditionally Periodic Motions under a Small Change of the Hamiltonian Function, *Dokl. Akad. Nauk.* 165 (1965) pp. 1245-1248.

Meshalkin, L. D.

[1] A Case of Isomorphism of Bernouilli Schemes, *Dokl. Akad. Nauk.* 128 (1959) pp. 41-44.

Milnor, J.

[1] Morse Theory, *Ann. Math. Studies,* Princeton (1961).

Moser, J.

[1] On Invariant Curves of Area-Preserving Mappings of an Annulus, *Nachr. Akad. Wiss. Göttingen* No. 1 (1962).

[2] On Invariant Surfaces and Almost Periodic Solutions for Ordinary Differential Equations, *Am. Math. Soc. Notices,* 12 No. 1 p. 124.

[3] New Aspect in the Theory of Stability of Hamiltonian Systems, *Commun. Pure Appl. Math.* 11 (1958) pp. 81-114.

[4] A New Technique for the Construction of Solutions of Non-Linear Differential Equations, *Proc. Natl. Acad. Sci. U. S.* 47 (1961) pp. 1824-1831.

[5] On Theory of Quasi-Periodic Motion, *Siam Review* (1966).

[6] A Rapidly Convergent Iteration Method and Non-Linear Partial Differential Equations I, II, *Annali Della Scuola Normale Superiore di Pisa,* Serie III 20 Fasc. II, III (1966) pp. 265-315; pp. 499-535.

[7] Convergent Series Expansions for Quasi-Periodic Motions, *Math. Annalen,* 169 (1967) pp. 136-176.

[8] On a Theorem of V. Anosov (in press).

Moser, J. and Jeffreys, W. H.

[9] Quasi-Periodic Solutions for the Three-Body Problem, *Astron. J.* 71 No. 7 (1966) pp. 568-578.

Nemytskii, V. and Stepanov, V. V.

[1] Qualitative theory of Differential Equations, Princeton (1960).

Ochozimski, D. E., Sarychev, V. A., Zlatoustov, V. A., and Torzevsky, A. P.

[1] Etude des oscillations d'un satellite dans le plan d'une orbite elliptique, *Kosmicheskie Issledovania* 2 No. 5 (1964) pp. 657-666.

Peixoto, M. H.

[1] On Structural Stability, *Ann. Math.*, Series 2, t. 69 (1959) pp. 199-222.

Poincaré H.

[1] Oeuvres complètes, 1 pp. 3-221.

[2] Les méthodes nouvelles de la mécanique céleste, 3. Gauthier-Villars, Paris (1899).

[3] Sur un théorème de géométrie, *Rendiconti circolo Mathematica di Palermo*, t. 33 (1912) pp. 375-407.

Polya, G. and Szegö, G.

[1] Aufgaben und Lehrsaetze aus der Analysis, 1, 2nd ed. Springer, Berlin (1954).

Rohlin, V. A.

[1] In General a Measure-Preserving Transformation Is Not Mixing, *Dokl. Akad. Nauk.* 13 (1949) pp. 329-340.

[2] On Endomorphisms of Compact Commutative Groups, *Izvestia Math. Nauk.* 13 (1949) pp. 329-340.

[3] On the Fundamental Ideas of Measure Theory, *Mat. Sbornik (N. S.)* 25, 67 (1949) pp. 107-150. [*Transl. Am. Math. Soc.* 1, 10 (1962) pp. 1-54.]

[4] Exact Endomorphisms of Lebesgue Spaces, *Izvestia Akad. Nauk.* 25 (1961) pp. 499-530. [*Transl. Amer. Math. Soc.* 2, 39 (1964) pp. 1-36.]

Rohlin, V. A. and Sinai, Ya.

[5] Construction and Properties of Invariant Measurable Partitions, *Dokl. Akad. Nauk.* 141 No. 6 (1961) pp. 1038-1041. [*Sov. Math. Dokl.* 2 No. 6 (1961) pp. 1611-1614.]

Schwartzman, S.

[1] Asymptotic Cycles, *Ann. Math.* 66 No. 2 (1957) pp. 270-284.

Siegel, C. L.

[1] Iterations of Analytical Function, *Ann. Math.* 43 (1942) pp. 607-612.

[2] Vorlesungen über Himmelsmechanik, Springer, Berlin (1956).

[3] Uber die Existenz einer Normalform analytischer Hamiltonsche Differentialgleichungen in der Nähe einer Gleichgewichtslösung, *Math. Annalen.* 128 (1954) pp. 144-170.

Sinai, Ya.

[1] The Central Limit Theorem for Geodesic Flows on Manifolds of Constant Negative Curvature, *Dokl. Akad. Nauk.* 133 (1960) pp. 1303-1306. [*Sov. Math. Dokl.* 1 No. 4 (1961) pp. 983-987.]

[2] Properties of Spectra of Ergodic Dynamical Systems, *Dokl. Akad. Nauk.* 150 (1963) pp. 1235-1237. [*Sov. Math. Dokl.* 4 No. 3 (1963) pp. 875-877.]

[3] Some Remarks on the Spectral Properties of Ergodic Dynamical Systems, *Usp. Mat. Nauk.* 18 No. 5 (1963) pp. 41-54. [*Russian Math. Surveys* 18 No. 5 (1963) pp. 37-51.]

[4] On the Foundations of the Ergodic Hypothesis for a Dynamical System of Statistical Mechanics, *Dokl. Akad. Nauk.* 153 No. 6 (1963). [*Sov. Math. Dokl.* 4 No. 6 (1963) pp. 1818-1822.]

[5] *Vestnik Moscovskovo Gosudrastvennovo Universitata.* Series Math. No. 5 (1962).

[6] Dynamical Systems with Countably Lebesgue Spectra, *Izvestia Math. Nauk.* 25 (1961) pp. 899-924. [*Transl. Amer Math. Soc.*, Series 2, 39 (1961) pp. 83-110.]

[7] On the Concept of Entropy of a Dynamical System, *Dokl. Akad. Nauk.* 124 (1959) pp. 768-771. [*Math. Review* 21 No. 2036a.]

[8] Letter to the Editor, *Usp. Math. Nauk.* 20 No. 4 (124) (1965) p. 232.

[9] A Weak Isomorphism of Transformations with an Invariant Measure, *Dokl. Akad. Nauk.* 147 (1962) pp. 797-800. [*Sov. Math. Dokl.* 3 (1962) pp. 1725-1729.]

[10] Geodesic Flows on Compact Surfaces of Negative Curvature, *Dokl. Akad. Nauk.* 136 (1961) p. 549. [*Sov. Math. Dokl.* 2 No. 1 (1961) pp. 106-109.]

[11] Dynamical Systems with Countably Multiple Lebesgue Spectra II. *Izvestia Math. Nauk.* 30 No. 1 (1966) pp. 15-68.

Sitnikov, K.

[1] The Existence of Oscillatory Motions in the Three-Body Problem, *Dokl. Akad. Nauk.* 133 No. 2 (1960) pp. 303-306. [*Sov. Phys. Dokl.* 5 (1961) pp. 647-650.]

Slater, N. B.

[1] Distribution Problems and Physical Applications, *Compositio Mathematica,* 16 fasc. 1.2 (1963) pp. 176-183.

Smale, S.

[1] Dynamical Systems and the Topological Conjugacy Problem for Diffeomorphisms, *Proc. Int. Congress Math.* (1962) pp. 490-496.

[2] Structurally Stable Systems Are Not Dense, *Am. J. Math.* 88 (1966) pp. 491-496.
[3] Differentiable Dynamical Systems, *Bull. Am. Math. Soc.* 73 (1967) pp. 747-817.

Stepin, A. M.
[1] On the Approximation of Dynamical Systems by Periodic Mappings and Its Spectrum, *Funkz. Analys i ego Prilojenija,* Moscos 1 No. 2 (1967).

Weyl, H.
[1] Sur une application de la théorie des nombres à la mécanique statistique, *Enseignement Math.* 16 (1914) pp. 455-467.
[2] Über die Gleichveirteiligung von Zahlen mod. 1, *Math. Annalen* 77 (1916) pp. 313-352.
[3] Selecta Hermann Weyl, Bassel-Stuttgart (1956) pp. 11-147.
[4] Mean Motion I, *Am. J. Math.* 60 (1938) pp. 889-896.
[5] Mean Motion II, *Am. J. Math.* 61 (1939) pp. 143-148.

Wintner, A.
[1] Upon a Statistical Method in the Theory of Diophantine Approximations, *Am. J. Math.* 55 (1933) pp. 309-331.

Yaglom, A. M. and Yaglom, I. M.
[1] Probabilités et informations, Dunod (Paris).

INDEX

fold and staple

No Postage
Necessary
if Mailed in the
United States

BUSINESS REPLY MAIL
FIRST CLASS PERMIT NO. 828 REDWOOD CITY, CA 94065

Postage will be paid by Addressee:

ADDISON-WESLEY PUBLISHING COMPANY, INC.®

Advanced Book Program
390 Bridge Parkway, Suite 202
Redwood City, CA 94065-1522